Lecture Notes in Mathematics

History of Mathematics Subseries

Volume 2261

Series Editor

Patrick Popescu-Pampu, CNRS, UMR 8524 - Laboratoire Paul Painlevé, Université de Lille, Lille, France

Providing captivating insights into facets of the recent history of mathematics, the volumes in this subseries of LNM explore interesting developments of the past 200 years or so of research in this science. Their aim is to emphasize the evolution of its intellectual discourse, the emergence of new concepts and problems, the groundbreaking innovations, the human interactions, and the surrounding events that all contributed to weave the backdrop of today's research and teaching in mathematics. These high-level and largely informal accounts will be of interest to researchers and graduate students in the mathematical sciences, in the history or the philosophy of mathematics, and to anyone who seeks to understand the historical growth of the discipline.

More information about this subseries at http://www.springer.com/series/8909

Martin Charles Golumbic • André Sainte-Laguë

The Zeroth Book
of Graph Theory

An Annotated Translation of Les Réseaux
(ou Graphes)–André Sainte-Laguë (1926)

 Springer

Martin Charles Golumbic
Department of Computer Science
University of Haifa
Haifa, Israel

André Sainte-Laguë
Conservatoire National des Arts et Métiers
Paris, France

Comments on the manuscript from mathematicians and other professional researchers are welcome. They should be sent to Martin Charles Golumbic, golumbic@cs.haifa.ac.il, Department of Computer Science, University of Haifa, Haifa, Israel.

ISSN 0075-8434   ISSN 1617-9692   (electronic)
Lecture Notes in Mathematics
ISSN 2193-1771   ISSN 2625-7157   (electronic)
History of Mathematics Subseries
ISBN 978-3-030-61419-5   ISBN 978-3-030-61420-1   (eBook)
https://doi.org/10.1007/978-3-030-61420-1

Mathematics Subject Classification (2020): 05C, 68

This Springer imprint is published by the registered company Springer Nature Switzerland AG
The registered company address is: Gewerbestrasse 11, 6330 Cham, Switzerland

Dedicated to the memory of André Sainte-Laguë's colleague, friend, and coauthor

Guy Iliovici

who was deported and murdered at Auschwitz by the Nazis and their collaborators.

Foreword

The appearance in 1926 of *Les Réseaux* (*ou Graphes*) was an important milestone in the historical development of graphs and their applications, from its first introduction by Euler (1735) with his famous problem on the bridges of Königsberg, to some early applications in chemical theory with the graphs of organic molecules. Most of the research in graph theory in the beginning of the last century still came from mathematical games and puzzles, but in his introduction (§2), Sainte-Laguë seemed to be convinced of the great potential of applications of graph theory.

Sainte-Laguë's book was the first textbook devoted to the study of graphs, containing most of what was known at his time, as for example, the four color problem. In many aspects his book is very modern, by its subject, its presentation, and with many examples. It is interesting that many important notions had already been defined: trees, centers, chains, cycles, Eulerian cycles, Hamiltonian cycles, etc., although some of the notations are not the ones we use today.

Sainte-Laguë, through Édouard Lucas's books *Récréations Mathématiques*, knew about depth-first search that had been introduced to deal with searching labyrinths by Trémaux (1882) and Tarry (1895). Euler and many others after him developed what they called Analysis situs (Géométrie de situation in French). Lucas and Sainte-Laguë introduced the word *graphe* to the French community, describing this new domain in order to extract it from geometry and metrics.

The translation of *Les Réseaux* (*ou Graphes*) is excellent. It is not easy to translate such a text, since some definitions are obsolete and others incomplete or even false. Golumbic has succeeded in keeping as close as possible to the original text. His added footnotes, remarks, clarifications, corrections, and commentaries will be useful to the reader. I am happy to see this historically important work published in this English version.

Many of my students in computer science network theory are always surprised when I cite Sainte-Laguë's book. How could something so modern be so ancient?

Michel Habib
Université de Paris

Preface

The making of this book

In 2011, while on sabbatical as a visiting professor at Columbia University, I first considered taking on the translation of *Les Réseaux (ou Graphes)*, which I had not known about previously. I also knew nothing about its author, André Sainte-Laguë, but I was attracted by the challenge of providing an English translation. A translation would serve the research community by drawing attention to this seminal monograph, more than 90 years after its first appearance. Moreover, it could be a valuable source for the future, by extensively annotating its ideas and fundamental principles.

Writing this book has turned out to be a worthwhile and interesting project, completing a circle of life. As a doctoral student at Columbia decades earlier, my own career was greatly influenced by the book *Graphs and Hypergraphs* by Claude Berge, and especially by his work on perfect graphs. So it was quite natural that after meeting Berge at a mathematics conference, I decided to spend my postdoctoral year in Paris. I attended Berge's seminars at Jussieu and his weekly informal problem sessions at Boulevard Raspail, while continuing my own research and starting the writing of my first book, *Algorithmic Graph Theory and Perfect Graphs*, for which Claude wrote a Foreword.

Thus, in many ways, I consider myself to have become, in my own way, a non-French member of what I might call the Berge School of Graph Theory. And who would have guessed—Claude Berge was born in 1926 in Paris, coinciding with the publication of *Les Réseaux (ou Graphes)*.

Martin Charles Golumbic
Haifa, Israel

MÉMORIAL

DES

SCIENCES MATHÉMATIQUES

PUBLIÉ SOUS LE PATRONAGE DE

L'ACADÉMIE DES SCIENCES DE PARIS

DES ACADÉMIES DE BELGRADE, BRUXELLES, BUCAREST, COÏMBRE, CRACOVIE, KIEW,
MADRID, PRAGUE, ROME, STOCKHOLM (FONDATION MITTAG-LEFFLER), ETC.,
DE LA SOCIÉTÉ MATHÉMATIQUE DE FRANCE, AVEC LA COLLABORATION DE NOMBREUX SAVANTS.

DIRECTEUR :

Henri VILLAT

Correspondant de l'Académie des Sciences de Paris,
Professeur à l'Université de Strasbourg.

FASCICULE XVIII

Les Réseaux (ou graphes)

Par M. A. SAINTE-LAGUË

Professeur au Lycée Carnot.

PARIS

GAUTHIER-VILLARS ET Cⁱᵉ, ÉDITEURS

LIBRAIRES DU BUREAU DES LONGITUDES, DE L'ÉCOLE POLYTECHNIQUE

Quai des Grands-Augustins, 55

1926

Title page of the original 1926 version by André Sainte-Laguë.

Contents

Tracing the topics in Les Réseaux (ou Graphes)

Martin Charles Golumbic

In 1926, André Sainte-Laguë published a 65-page monograph entitled *Les Réseaux* (*ou Graphes*)[1] which Harald Gropp has called the 'zeroth' book on graph theory [248]. The monograph appeared only in French, but Gropp published a number of papers in English between 1993 and 2003 discussing historical aspects and giving a modern interpretation of several topics. He also expressed his hope that an English translation might sometime be available to the mathematics community.

In the 10 years following the appearance of *Les Réseaux* (*ou Graphes*), the development of graph theory continued, culminating in the publication of the first full book on the theory of finite and infinite graphs in 1936 by Dénes Kőnig [268]. This remained the only well-known book for 20 years, until 1958 when Claude Berge published his book on the theory and applications of graphs [228]. By 1960, graph theory had emerged as a significant mathematical discipline of its own.

Les Réseaux (*ou Graphes*) provides historical testimony and a reminder of the excellent research in graph theory that had already been conducted at that time. Many of the concepts treated were still in their infancy, and other more developed notions offered a wealth of mathematical challenges. It is an important work, yet unfamiliar to most graph theorists.

By necessity, this book goes far beyond a mere straightforward translation. Dissecting, understanding, and converting concepts and terminology written a century ago, raised many difficult yet interesting challenges for presenting the work to a modern audience. We have annotated the text with over one hundred explanatory footnotes, provided remarks and commentaries on the original French, additional references, as well as supplementary comments by Alain Hertz, Myriam Preissmann and Robin Wilson. A casual reader, less interested in the specifics of the translation, may still appreciate these commentaries, in addition to the biography of André Sainte-Laguë that follows the translation.

---

[1] André Sainte-Laguë, Les Réseaux (ou Graphes). (Mémorial des Sciences Mathématiques, No. 18.) Paris, Gauthier-Villars, 1926. Available at http://eudml.org/doc/192551

© Springer Nature Switzerland AG 2021
M. C. Golumbic, A. Sainte-Laguë, *The Zeroth Book of Graph Theory*, Lecture Notes in Mathematics 2261, https://doi.org/10.1007/978-3-030-61420-1_1

A snapshot from 1926

Les Réseaux (*ou Graphes*) provides an important 'snapshot' of what was known about graph theory in 1926. Michel Habib describes it as a milestone on the road of graph theory and algorithms. André Sainte-Laguë was clearly influenced by the volumes of Édouard Lucas, Walter William Rouse Ball, Henri Poincaré, and others from the 1880s and 1890s, on what was then often considered to be recreational mathematics. With the perspective of more than a century of development, we now see this as a foundational period of combinatorics.

On the algorithmic side, graph traversal and depth-first search were already used by Tarry and Trémaux in the 1880s, and their very nice application to the computation of an Eulerian chain was given by Fleury in 1883.

Sainte-Laguë also prophetically understood that applications would play a future role in the development of the discipline. But it is doubtful that anyone in 1926 with a passion for mathematics, science, or technology, nor a forward-looking Jules Verne, could have imagined the *Voyages Extraordinaires* (Extraordinary Journeys) that graph theory has taken since. Not even H. G. Wells.

There are rather few citations of *Les Réseaux* (*ou Graphes*) in the literature. Kőnig cites it in his 1936 book [268] several times *"and especially mentions his bibliographical list which was useful to him ... to investigate the development of early graph theory in the French mathematical literature."* [248]. An interesting historical perspective about Kőnig and his book can be found in [299].

W. T. Tutte wrote in [298] (translated from the original French):

> At Cambridge University, I found a work of Sainte-Laguë entitled *Les Réseaux* (*ou Graphes*). There is a proof of Petersen's theorem. I read. I understood. I filled the gaps. I even made a small improvement in the result of the text. 'Look at you,' I said to myself, 'You can work on networks. Perhaps the theory of graphs will be your research topic in the future!'

Sainte-Laguë's manuscript is listed in the biography of the historical book by Biggs, Lloyd and Wilson [230] without further reference.

Sainte-Laguë's book begins its journey by introducing most of the basic definitions of graph theoretic notions commonly known at the time, using terminology that was both standard and non-standard in French. It then moves on to a variety of topics in graph theory at that time. He also published a short sequel [285], beginning by discussing the four color problem in more detail, and then moving on to graphs and games.

Translation is always a daunting task

Translating a mathematical text written over 90 years ago can be a formidable task. I have attempted to make the translation faithful to the original, yet occasionally a slightly freer translation is given to make it easier to read. Much of the translation uses terminology that is standard today, and which is different from that used by Sainte-Laguë. This decision was taken to help you to understand the concepts presented, thus serving as a basis of comparison. As an aid, a Glossary of translated terms has been added at the end of the book.

In some instances, however, there were no obvious translations of terms used by Sainte-Laguë. For example, in (§6), his use of the term *isthme* is slightly more restrictive than today's usage of the term *isthmus*—namely, he did not consider a pendant edge (*impasse*) to be an *isthme*. Another example is the French word *feuille*, meaning a leaf, yet the current usage of the term *leaf* in graph theory is not the concept that Sainte-Laguë had in mind. We translate *feuille* as a *leaf component* whose definition is a subgraph obtained by deleting an isthmus, but which itself contains no isthmus. This translation was chosen to be analogous to the definition of a leaf block (*membre*) of a graph—namely, a subgraph that has no cut vertices and has only one vertex in common with the rest of the graph.

To maintain fidelity to the literal meaning of the presentation, illustrations (numbered figures) are the originals from his manuscript. His section numbering and format are maintained, to assist those who may read the French and English versions in tandem. I have also retained the mathematical notation used by Sainte-Laguë, even when it varies from section to section. For example, a graph (*reseaux*) is usually denoted by (R), including the parentheses, but in Chapter V, he switches to (ρ).

The process of proofreading the original French version also uncovered mathematical discrepancies and errors. These needed to be corrected in the translation—or sometimes had to be left alone with just a footnote.

The Bibliography begins with an edited version of the original appearing in the work. This is followed by further references cited in the commentaries and footnotes. The Glossary provides the French and English terms used in the book together with the section where they first appear.

Commentary

Sainte-Laguë had a reputation for integrating humor and personality into his presentations, first in his years of high school teaching, and later at the university attracting a thousand listeners at a time. Such elegance, however, is not reflected in this 1926 opus. There are no jokes, allusions, humorous mathematical bits, or fun puzzles to spice up this work. Nonetheless, I have taken the liberty of adding a few interesting footnotes and commentaries at the end of most chapters. For example, how many young readers today can describe a strip of postage stamps without seeing an example, or appreciate the art of a magic square?

Finally, I have written a short biography of André Sainte-Laguë to give a glimpse into his fascinating mathematical and non-mathematical career. To understand the importance of this work, it is worthwhile to see it in the context of his entire career.

Networks (or Graphs) — André Sainte-Laguë (1926)

André Sainte-Laguë among his students at the
Lycée Janson-de-Sailly (Paris) in 1924

Image: courtesy of the Central Library of the Conservatoire National
des Arts et Métiers (CNAM), Paris, [236]

I Introduction and definitions

1. **Generalities**. — Topological[2] problems can be viewed from two different perspectives. Either one tries to study continuous deformations that allow one to pass from one curve or surface to a different curve or surface, and this is what is done in particular in *Analysis situs* [13, 20, 117, 118, 119, 120], or one excludes completely any idea of measure and one considers only relative placement of the objects given. In the second case, which will be the only one treated here, one asks questions of two different types:

a. Is it possible to place given objects next to each other so that one obtains a particular given configuration?

b. If such a placement is possible, in how many ways can it be done?

The first genre of questions constitutes *par excellence* the research in topology, free of all metric considerations, as we have already said. The second genre contains additionally the problems of enumerative geometry and uses properties of integers, numerical sequences, combinatorial analysis. The answers given are often only of empirical nature: in spite of important progress achieved in topology, this field does not yet form a very coherent theory and the results that constitute its substance are a bit disparate. It is only recently that systematic studies, related mostly to the "four color problem", were undertaken. It seems that with a little more effort the results could be infinitely superior to the present ones.

2. This lack of activity on the part of researchers is quite regrettable. In fact, the applications are already very interesting and will become without doubt even more so. One should not think that topology can only be applied to mathematical curiosities such as the "Königsberg bridges problem" [25] or the four color problem[3], a theorem[4] which, it must be said, seems to be as mysterious as certain properties of prime numbers. Topology is not only used in the study of many *games*, but also it is a foundation for numerous scientific research. Be it *invariants* [34], *determinants* [34], *resultants* of two equations [28], *analytic forms* [17], *arithmetic* [14, 15], *groups*

---

[2] *Géométrie de situation*, in the terminology of *S.-L.*

[3] *S.-L.*: We will address this question in a booklet *Géométrie de Situation et Jeux*, in preparation, *Mémorial des Sciences Mathématiques*. [Published [285] as No. 41 (1929). Available at http://eudml.org/doc/192569]

[4] Still a conjecture until 1976. Robin Wilson's fascinating book, *Four Colors Suffice* [304], relates the story of how the famous 120 year old four color problem was solved.

© Springer Nature Switzerland AG 2021
M. C. Golumbic, A. Sainte-Laguë, *The Zeroth Book of Graph Theory*,
Lecture Notes in Mathematics 2261, https://doi.org/10.1007/978-3-030-61420-1_2

of substitutions [38], or *permutations* [37], or even *graphic statics* [19], topological considerations play a role.

But it seems that main applications of this branch of mathematics will be in *Physical Chemistry* or in *Chemistry*. It suffices to browse modern research papers on the composition of matter or the structure of crystals [33] to become aware of that. A whole branch of Organic Chemistry [17, 18, 21, 25, 27, 38] seems to depend on graph theory, and the study of paraffins [17] already has given rise to interesting research.

3. Topology, as we see it, is very close to formal logic in that it is based on very simple basic notions. Its starting point is the fundamental notion of *associated* or *non-associated* elements. Thus, for example, when moving a piece on a chessboard (§67), two squares would be considered associated if a piece can be moved from one to another in a single turn.

This idea of association or non-association of distinct elements can be found in many research papers, but is generally in a hidden form. Nevertheless, de Polignac [22] has already considered *n* elements that are pairwise *varying* or *permanent*.

The simplest and often the most convenient way to study properties of a list of elements $A, B, \ldots$ which are pairwise associated or non-associated is to represent them by points: two associated elements are, by convention, connected by a line, and non-associated elements are not connected. We already find such representations in Cayley [16].

In this way, one is brought to a formalization of a notion of *graph* to which we will reduce systematically all related questions.

4. **Definitions**. — A *network* or *graph*, after Sainte-Laguë [36], who adopted the notion already introduced by Lucas [32], is a set of points, crossroads or *vertices* joined by lines or paths which are the *edges* of the graph. The form of the edges is not important, what matters is only whether two vertices A, B are connected by one or several edges.

Two graphs are *homeomorphic*[5] if there is a bijective correspondence between their vertices, on the one hand, and between their edges, on the other hand. Such graphs are in practice considered as identical.

An edge AB always goes between two vertices A, B, which are its *endpoints*[6] and cannot contain any other vertex. In certain cases, when we cut a graph in two, we might have to consider a *half-edge* which only has one end. An edge AB is called *simple*, *double*, *triple*, $\ldots$ if there are 1, 2, 3, $\ldots$ edges joining the same endpoints A, B.

A *pendant edge*[7] is an edge that connects A to B, when A is not connected to any other vertex but B. If there are 1, 2, 3, $\ldots$ edges AB, the pendant edge is *simple*,

---

[5] The term used nowadays is *isomorphic*.

[6] *extrémités*, in the terminology of S.-L.

[7] *impasse*, in the terminology of S.-L.—its non-mathematical meaning is a *dead-end*.

double, triple, ... The vertex A is called a *pendant vertex.*[8] A *loop* is an edge whose endpoints are joined together at one vertex A. If there are 1, 2, 3, . . . such edges, the loop is called *simple, double, triple, ...* We call A a *loop vertex.* If A does not belong to any edge, except perhaps some loops, it is an *isolated vertex.* It is a *degree 2 vertex*[9] if only two edges end in it. A vertex is of *degree p* if it has p edges ending in it, each multiple edge counted according to its multiplicity, and each loop counted twice. A vertex is *even* or *odd* depending on the parity of p.

A graph is *symmetric* if it is homeomorphic to a graph in which vertices and edges are pairwise symmetric with respect to a center, an axis, or a plane of symmetry. We call vertices of the first graph *symmetric vertices* if they correspond to vertices that are symmetric to each other in the second graph. A and B are *indiscernible vertices* if the graph that contains them is homeomorphic to itself, where A corresponds to B and B corresponds to A. Two symmetric vertices are indiscernible.

5. A *chain* is a set of n edges $AB, BC, . . . , KL$ in which the same edge does not appear twice. A vertex of a chain is a *crossing vertex*[10] if it is repeated several times in a chain. So, it is in the chain an even number of times[11] except for the end vertices A and L. If A and L are equal, the chain is called a *cycle.*[12] A chain is *even* or *odd* depending on the parity of n.

We can fix the direction of a chain. This gives rise to the notion of *oriented* chains, edges, and cycles. An edge that can be followed only in one direction is called *unicursal*; if it can be followed in two directions, then it is called *bicursal* [34, 99] (§44).

A chain or a cycle without crossing vertices is called *simple*[13]; it is *complete*[14] if it goes through all the vertices of the graph (§24). A graph is *traceable*[15] if it admits a complete chain, and *circularly traceable*[16] or *Hamiltonian* if it admits a complete cycle (§57). A graph for which there exists a chain or a cycle that includes all the edges is called an *entrelacement (interlacing)* (§16). It is called an *Eulerian chain*[17] in the first case, and an *Eulerian cycle*[18] in the second case.

A graph is *connected* if there is always a chain from A to B, for arbitrarily chosen vertices. It has 2, 3, . . . *connected components* if it consists of 2, 3, . . . connected graphs with no pairwise common vertices or edges. A graph is a *subgraph* of another

---

[8] *sommet d'impasse*, in the terminology of *S.-L.*

[9] *sommet de passage*, in the terminology of *S.-L.*

[10] *sommet de croisement*, in the terminology of *S.-L.*

[11] *S.-L.* clearly means that it appears in an even number (≥ 4) of edges.

[12] *circuit*, in the terminology of *S.-L.*, who uses the term *cycle* for an oriented cycle.

[13] *circulaire*, in the terminology of *S.-L.*

[14] *Hamiltonian*, in today's terminology.

[15] *aligné*, in the terminology of *S.-L.*

[16] *cerclé*, in the terminology of *S.-L.*

[17] *entrelacement ouvert*

[18] *entrelacement fermé*

graph if all of its vertices and edges are among the vertices and edges of the other graph, which is called the *containing graph*.

6. A *cut-vertex*[19] is a vertex A of a graph for which the remaining vertices can be partitioned into two groups P and Q, such that every path from a vertex of P to a vertex of Q necessarily passes through A (§20). A *leaf block* of a graph is a subgraph with no cut vertices which has only one vertex in common with the rest of the graph. The number of leaf blocks cannot be equal to 1. [20]

An *isthmus* is a non-pendant edge AB such that one can divide all other vertices into two groups P and Q in such a way that every chain from a vertex of P to a vertex of Q contains the isthmus;[21] if one of the obtained subgraphs contains no an isthmus, this subgraph is called a *leaf component*[22] [34]. The number of leaf components cannot be 1.

A *tree* is a connected graph in which there is always a unique chain between any two vertices (§9, §49). It has no loops, multiple edges, or degree 2 vertices.[23] It is called a *star* if it contains no cut-edges; in this case, all edges are pendant edges. [The degree of the star is the number of edges.]

7. A graph is *finite* if the number of vertices and the number of edges are finite. Otherwise, it is *infinite*. The *order* of a finite graph is the number of its vertices. A graph is *normal* if it is *finite*, *connected*, and has none of the following elements (which we will call *singularities*): multiple edge, pendant edge, loop, degree 2 vertex, cut-vertex, or isthmus. It is *complete* if it has edges that join all vertices pairwise (§27).

A normal graph is *polygonal* if, given its order n, it is homeomorphic to a graph drawn on a plane that can be made to coincide with itself after a rotation of $\frac{2\pi}{n}$ (§28). Any two vertices are then indiscernible (§4). A normal graph is *semi-polygonal* if any two vertices are indiscernible. This is the case for the non-polygonal graph formed by the vertices and edges of a cube.

A graph is *regular* and of *degree p* if all of its vertices are of degree p. It is *cubic* if $p = 3$, *quartic* if $p = 4$.

The *chromatic number*[24] of a graph is the minimal number of groups into which we can divide the vertices so that two vertices of the same group are not joined by an edge. A graph with chromatic number 2, 3, 4, . . . is *bipartite* (§52, §55), *tripartite*,

---

[19] *articulation*, in the terminology of *S.-L.*

[20] A leaf block is a *membre*, in the terminology of *S.-L.*; in current terminology, a maximal subgraph that contains no cut-vertex is called a *block*.

[21] *isthme*, also known as a *cut-edge* or a *bridge*, however, today one generally regards a pendant edge to be an isthmus as well. So we must be careful not to confuse *S.-L.*'s notion with today's accepted common usage of the word *isthmus* in graph theory.

[22] *feuille*, in the terminology of *S.-L.*

[23] In its modern usage, the term *tree* omits the requirement that there are no degree 2 vertices— *sommets de passage* in the terminology of *S.-L.*

[24] *rang*, in the terminology of *S.-L.*

tetrapartite, ... The *edge-chromatic number*[25] of a graph is the minimal number of groups into which we can divide all edges so that two edges of the same group never have a common vertex.

A graph has *genus* 0, 1, 2, ... if it is homeomorphic to a graph whose edges can be drawn on a surface of genus 0, 1, 2, ..., so that two edges never cross each other, and consequently have no common points apart from their ends. A graph of genus 0 is called *spherical*: it can be drawn on a sphere or on a plane.[26] In this situation, the term "edges" is justified by analogy with the corresponding terminology for polyhedra. Similarly, we call a *face* any portion of a sphere that is surrounded by edges and has no edge in its interior. Juxtaposing all faces, we get the entire surface of the sphere. If we look at the representation of the graph by stereographic projection on the plane, making sure that there are no infinite edges, then one of the faces contains infinite parts of the plane: this is the *exterior face* or the *sea*, as opposed to the *interior faces* or the *continent*. A spherical graph is *rhombic* if all faces are quadrilateral, and *triangulated* if they are triangles. It is easy to give similar definitions for a *toroidal* graph, etc.

If C is the number of edges of a graph, and S is the number of vertices, then the *index* of the graph is the integer $\omega = C - S + 1$ (§15, §49).

8. Two normal graphs, or compositions of a finite number of normal graphs, are *associated*[27] (§52) if there is a bijective correspondence between their vertices so that if two vertices are connected by an edge in one, then they are not connected by an edge in the other, and vice versa. If two normal graphs are associated, each one is *associable*. A *self-complementary* graph is a graph that is homeomorphic to its associated graph.

If we divide the vertices of a graph into two groups, in different ways, so that each group contains at least two vertices, then the minimum number of edges that join vertices of one group to the vertices on the other group is the *strength*[28] of the graph. If there is an isthmus, then the strength of the graph is 1, and vice versa.

The *distance* between two vertices of a normal graph is the minimum number of edges in a chain that connects these two vertices. It is 1 for vertices joined by an edge. If we take a vertex A to be of *level* 0,[29] all vertices at distance 1 from A will be of level 1, at distance 2 of level 2, ... The maximum level is the *height*[30] of the graph relative to A; the maximum and minimum values of the height are the *diameter*[31] and *radius*[32], which are the two *dimensions* of the graph. The graph is

---

[25] *classe*, in the terminology of S.-L.

[26] Today we would use the term *planar*.

[27] This term really means *complementary* in modern terminology, but is defined here for (finite unions of) normal graphs, and the complement is required to be in the class of graphs too.

[28] *puissance*, in the terminology of S.-L.

[29] *cote*, in the terminology of S.-L.

[30] *hauteur*, known today as the *eccentricity* of A.

[31] *longueur*, in the terminology of S.-L.

[32] *largeur*, in the terminology of S.-L.

balanced[33] if these two are equal, [and their common value is called the *dimension* of the graph.]

In a normal graph, a *maximum clique*[34] is a complete subgraph of order m, which is as big as possible; m is the *maximum clique number*[35] of the graph.

---

[33] *rond*, in the terminology of *S.-L.*

There is still no known characterization of graphs whose radius equals its diameter. Upper and lower bounds can be found in: Paul Erdős, János Pach, Richard Pollack and Zsolt Tuza, Radius, diameter, and minimum degree, *Journal of Combinatorial Theory* (B) 47 (1989), 73–79.

[34] *noyau*, in the terminology of *S.-L.*

[35] *grosseur*, in the terminology of *S.-L.*

Commentary. The terms *Géométrie de Situation* and *Analysis Situs* were in common use by mathematicians during the period when Sainte-Laguë wrote his manuscript. They are generally translated today as topology, a term that was not commonly used then. They embody the notion of "continuous deformations".[a]

Sainte-Laguë was influenced by his reading of the classical works of Poincaré, Lucas, and others. The Latin "Analysis Situs" was the title of an influential paper by Henri Poincaré, published in 1895. Over the following ten years, Poincaré published five supplements to the paper. "Analysis Situs" was also the title of a book by the Princeton mathematician Oswald Veblen, published in 1922, and based on his 1916 lectures at the Cambridge Colloquium of the American Mathematical Society. It is considered to be the first English-language textbook on topology, and served for many years as the standard reference. Its contents were based on the work of Henri Poincaré, as well as Veblen's own work with his former student and colleague, James W. Alexander. These papers provided the first systematic treatment of topology and revolutionized the subject.

Sainte-Laguë concluded section (§2) by saying that if somewhat greater efforts were made (*des efforts un peu plus grands*), then the results would very quickly (*très vite*) be infinitely superior to what they are today. Given the major progress of graph theory research witnessed during the 1930s, he was right. Moreover, after Dénes Kőnig's book [268] appeared in 1936, the field attracted more and more attention. By the 1960s, graph theory had caught on like wildfire—growing exponentially. The driving force was its close relationship with the emerging uses of graphs as models and for the optimization of networks.

With this in mind, Section (§2) could have been called, "The potential for applications of graph theory". Sainte-Laguë saw this potential, as perhaps others did at that time. But they could not have envisioned how graph theory applications have invaded virtually every discipline in the mathematical, computational, and social sciences. He mentions the Four Color Problem and the Königsberg Bridge Problem, which he refers to as "mathematical curiosities". Indeed, most work in combinatorics was regarded as recreational mathematics at that time (and for a long time afterwards too), totally unjustified considering the works of Euler, Kirchhoff, Cayley, and others. But Sainte-Laguë tells us that this field would become much more important for applications, citing several branches of games and chemistry. He saw the tip of what would become the *graph theory iceberg*.

---

[a] *Robin Wilson comments:* The word *topology* was introduced by Listing in 1838, and appears in the title of his 1847 book *Vorstudien zur Topologie*.

II Trees

9. **Chain covers of a tree**. — We use the term *rameau* for a pendant edge in a tree[36] (§6, §49) (Fig. 1), and each vertex which is not a pendant vertex is an *internal node*[37]. Following de Polignac [23], a *chain cover*[38] is a subset of chains such that

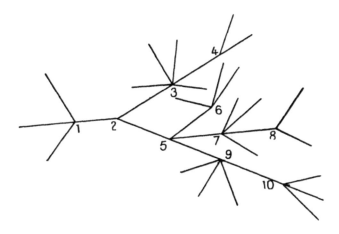

Fig. 1.

each edge belongs to exactly one chain. The first chain must contain two pendant edges, while others can [start or] end in an internal node.[39] Let $m_2, m_3, \ldots, m_p$ be,

---

[36] *S.-L.* could have continued to use his terminology *impasse* (dead-end) for a pendant edge, but for the special case of a tree, he chose the word *rameau* meaning a small branch or bough of a tree.

[37] *noeud*, in the terminology of *S.-L.*

[38] *traits*, in the terminology of *S.-L.*

[39] It is likely, in view of what appears below and in Section (§10), where *S.-L.* uses the same term *trait* for one chain of a chain cover, that he is thinking in terms of a construction procedure (unstated) for iteratively choosing chains. Moreover, each successive chain cannot be lengthened further.

© Springer Nature Switzerland AG 2021
M. C. Golumbic, A. Sainte-Laguë, *The Zeroth Book of Graph Theory*,
Lecture Notes in Mathematics 2261, https://doi.org/10.1007/978-3-030-61420-1_3

respectively, the numbers of internal nodes of even[40] degree: $4, 6, \ldots, 2p$, and n_1, $n_2, \ldots, n_q$ the numbers of internal nodes of odd[41] degree: $3, 5, \ldots, 2q + 1$. Let[42]

$$\sum_{i \geq 2} m_i = M, \qquad \sum_{j \geq 1} n_j = N, \qquad \sum_{i \geq 2} i\, m_i = P, \qquad \sum_{j \geq 1} j\, n_j = I.$$

The total number of internal nodes is thus $M + N$. We will find that if R is the number of pendant edges (*rameaux*) and C is the number of edges, then, after Sylvester [46], the *magnitude*[43] of the tree is

$$2C = n + \sum_{i \geq 2} 2i\, m_i + \sum_{j \geq 1} (2j + 1)n_j = R + 2P + 2I + N.$$

For a *star* (§6) of degree $2t$, the cardinality of a chain cover is t; if the degree is $2t + 1$, then this number is $t + 1$.[44] The cardinality T of a chain cover in a tree with $2, 3, \ldots$ internal nodes is given by

$$T = 1 + \sum_{i \geq 2} i\, m_i + \sum_{j \geq 1} (j + 1)n_j - \sum_{i \geq 2} m_i - \sum_{j \geq 1} n_j = 1 + P + I - M.$$

Therefore, *the cardinality T of a chain cover is independent of [the order in] the method used to make the list of them.* This number T is called the *base* of the tree.

The number R of pendant edges of a tree comprised of only n_j internal nodes of degree $2j + 1$ is $T + 1 + (j - 1)n_j$, and if there are only m_i internal nodes of degree $2i$, it is $T + 1 + (i - 1)m_i$. From this, we deduce

$$R = T + 1 + \sum_{i \geq 2} (i - 1)m_i + \sum_{j \geq 1} (j - 1)n_j = T + 1 + P + I - M - N,$$

from which two relations follow:

$$R = 2 + 2P + 2I - 2M - N, \qquad C = 1 + 2P + 2I - M.$$

The total number S of vertices of the tree is given by

$$S = R + M + N = 2 + 2P + 2I - M = C + 1.$$

---

[40] Notice that degree 2 is not listed here since, in (§6), a tree is defined as a graph without "*sommet de passage*"—that is, with no degree 2 vertices.

[41] *Corrected typo:* Corrected from the original $2, 5, \ldots$,

[42] *Corrected typo:* Indices of the summations have been corrected in this section from those of original version.

[43] The term *magnitude* is used here for the only time in the manuscript. In the formula, n is undefined; an easy algebraic computation leads to the first summation equaling $2P$ and the second summation equaling $2I + N$, so perhaps *S.-L.* intended that n should have been R, i.e., a typo in the original manuscript.

[44] This seems to confirm our speculation that *S.-L.* had in mind an unstated procedure of iteratively choosing chains of maximal length until all edges are covered exactly once.

10. **Tables**. — De Polignac [23] associates a *table* with each tree, the vertices being arbitrarily ordered, where each row represents a chain of the tree, with the exception of the pendant edges. Thus, the tree in Fig. 1 can be represented by one of the two tables:[45]

1 2 5 7 8	7 8
4 3 2	1 2 5 7
6 5 9 10	4 3 2
	6 5 9 10

The first table is called *irreducible*[46], while the second is *reducible*, because the number 7 is situated at the extremities of two chains that can be united into a single one.

Considering such tables helps us to understand the invariance of the number T. A table is irreducible if each row contains a unique number that is already written in the previous rows, a condition that is not invariant under permutations of rows.

11. **Centers**. — Jordan [44], and then Sylvester [46] and Cayley [39], introduced the notion of a *center* of a tree, which was later used by de Polignac [23] and Delannoy [42]. Start with an internal node A and follow a chain that necessarily ends with a pendant edge. Let m be the height of this chain (§8)—that is, the number of edges that it contains, including the final pendant edge.

If h and h' ($h' \geq h$) are the heights of the two chains of largest height starting at A, [each via a different neighbor of A], then they form a chain of length $h + h'$, the longest chain passing through A. [Now choose A such that the length $h + h'$ is longest possible in the graph.] If this maximum length is an even number $2m$, then the internal node O situated in the middle of this chain is the center of the tree. If the maximum length is $2m - 1$, then the internal nodes O and O', the endpoints of the edge in the middle of the path, are the two *centers* of the tree. This establishes that *a tree either has a single center, or has two centers, at the endpoints of a single edge.*

12. Trees with number of internal nodes fixed in advance. —

Finding the number $\varphi(n)$ of pairwise non-homeomorphic trees with n internal nodes is made easier through the notion of a center, considered as a point where the graph is cut into two. Each of the two halves constitutes a *branch*[47] of height m. The terms *arbre-racine* or *wurzelbaum* have been used in the same way [39].[48]

The computation of $\varphi(n)$ for branches of n nodes was considered only for the case in which the degree of each internal node is 2, 3, or 4, because the chemical

---

[45] among others.

[46] The definition of *irreducible* comes after an example, but it does *not* match the example. We are referred to de Polignac [23].

[47] *tige*, in the terminology of *S.-L.*

[48] Here *S.-L.* assumes that there is one center O—namely, that the longest chain is of even length $2m$. The tree is 'ripped apart' arbitrarily into two halves, each called a *branch*. Below, he considers only the case of trees with maximum degree 4, implying that within each branch the vertex O has degree 1, 2, or 3.

compositions of hydrogen and carbon atoms are of this kind [43], as in the case of paraffins C_nH_{2n+2} studied by Cayley [17, 39]. He found that the number $\varphi(n)$ satisfies the identity:[49]

$$[1 + \varphi(1)x + \varphi(2)x^2 + \ldots](1-x)(1-x^2)^{\varphi(1)}(1-x^3)^{\varphi(2)} \cdots \equiv 1.$$

Some inaccuracies have been found in the work of Cayley by Delannoy [42], and so we follow this latter author.

13. A branch, therefore, can have only 1, 2 or 3 edges originating from the cut-vertex. If it is of height m, then it has a chain of height m, while other chains can be of heights ranging from 1 to m. If $_\alpha t_n^m$ is the number of branches with α edges, n internal nodes, and height m, and if $T_n^m = {}_1t_n^m + {}_2t_n^m + {}_3t_n^m$ is the total number of branches with 1, 2, 3 initial edges, then we have

$$_1t_n^m = T_{n-1}^{m-1} \qquad \text{and then} \qquad _2t_n^m = \sum T_p^f T_{n-(p+1)}^{m-1}$$

p going from 1 to $n - (m+1)$ and f going from 0 to $m-1$. Finally, we have

$$_3t_n^m = \sum T_p^f T_q^s T_{n-(p+q+1)}^{m-1}$$

p going from 1 to $\frac{1}{2}(n-m-1)$, q going from p to $n-(m+p+1)$, and f and s going from 0 to $m-1$. Denoting by $_2a_n^m, _3a_n^m, _4a_n^m$ the number of trees with one center, of height m, with 2, 3, 4 . . . initial edges, we equally have

$$_2a_n^m = \sum T_p^{m-1} T_{n-p-1}^{m-1}$$

p going from m to the largest integer less than or equal to $\frac{1}{2}(n-1)$ and

$$_3a_n^m = \sum T_p^f T_q^{m-1} T_{n-p-q-1}^{m-1}$$

p going from 1 to $n - 2m - 1$, q going from m to the largest integer less than or equal to $\frac{1}{2}(n-p-1)$, f going from 0 to $m-1$, and finally

$$_4a_n^m = \sum T_p^f T_q^s T_r^{m-1} T_{n-p-q-r-1}$$

where p goes from 1 to the largest integer less than or equal to $\frac{1}{2}(n-2m-1)$, q goes from p to $n - 2m - p - 1$, r goes from m to the largest integer less than or equal to $\frac{1}{2}(n-p-q-1)$, and f and s go from 0 to $m-1$.

The number b_n^m of trees with two centers is given by the formula $b_n^m = {}_2a_{n+1}^{m+1}$, since a tree with one center (Fig. 2) can be made into a tree with two centers by removing the node A and joining B and C by an edge.

Delannoy gives as a conclusion the following numerical results:

---

[49] At the end of the 'identity', *S.-L.* used the congruence symbol $\equiv$.
Robin Wilson comments: This is made clear in Cayley's paper.

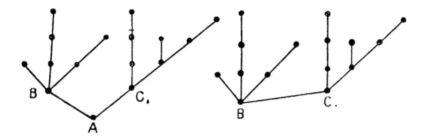

Fig. 2.

n	1	2	3	4	5	6	7	8	9	10	11	12	13
1 center	1	0	1	1	2	2	6	9	20	37	86	181	422
2 centers	0	1	0	1	1	3	3	9	15	38	73	174	380
total	1	1	1	2	3	5	9	18	35	75	159	355	802

14. The same author described a different method, in which, denoting by n', n'', n''' the number of internal nodes with 2, 3, 4 outgoing edges, and by r the number of pendant vertices, he established the relations

$$r = 2 + n'' + 2n''' \qquad \text{and} \qquad n' + n'' + n''' + r = n,$$

from which we deduce that

$$n' + 2n'' + 3n''' = n - 2.$$

For $n \leq 11$, we have the following empirical formulas for $\varphi(n)$, which was earlier defined (§12), for n even and odd:[50]

$$\varphi(2p) = 1 + (p - 1) + (p - 2)(p - 1) + \frac{p - 3}{3}(2p^2 + 3p - 20)$$
$$+ \frac{p - 4}{6}(2p^3 + 62p^2 - 537p + 1053) - (2p - 9)^2,$$
$$\varphi(2p + 1) = 1 + 2(p - 1) + (p - 2)(3p - 5) + \frac{p - 3}{3}(14p^2 - 66p + 82)$$
$$+ \frac{p - 4}{6}(18p^3 - 42p^2 - 471p + 1491) - 2[(2p - q)^2 + 1].$$

In these formulas, we must refrain from canceling out the terms, since negative factors would have to be replaced by 0.

15. **Index of a graph.** — The identity (§9) $S = C + 1$ shows that the *index* (§7, §49) of a tree is zero: $\omega = C - S + 1 = 0$.

---

[50] *Corrected typo:* The term $62p^2$ appeared incorrectly as $62p^3$ in the original.

We find that *in any connected graph, the index ω is a positive number or zero* [24, 41]. The index is zero only for trees.

Let $a_1, a_2, \ldots, a_C$ be distinct edges of a graph, considered as being *oriented* (§5). An oriented chain will be represented by

$$\sum_{i=1}^{i=C} \varepsilon_i a_i,$$

where ε is equal to $1, -1$, or 0 depending on whether the given edge a_i is taken in the positive or negative direction, or does not occur. For any directed cycle, we have the congruence

$$\gamma \equiv \sum_{i=1}^{i=C} \varepsilon_i a_i.$$

A subset of directed cycles is *linearly independent* if we cannot find integers m such that

$$\sum m\gamma \equiv 0.$$

We find, moreover, that in a graph of index ω, there are precisely ω linearly independent directed cycles for which any directed cycle can be represented as a linear combination of them. The case of loops is included in this statement.

Commentary. Sainte-Laguë was ahead of his time in the use of illustrative slides, pictures and films in his lectures at the Conservatoire National des Arts et Métiers (CNAM). This was particularly true for introductory courses, during the early 1930s, where he motivated the mathematics underpinning each application to be taught to an engineering audience.

Below is one of the actual classroom slides used by Sainte-Laguë to show the tree structure exhibited by sea corals.

Image: courtesy of the Central Library of the Conservatoire National des Arts et Métiers (CNAM), Paris, [236]

III Chains and cycles

16. Interlacings (Eulerian chains and cycles). — The question of whether a given graph admits a chain or cycle passing through every edge exactly once[51] was first asked by Euler [25], but must have been known before, as shown for instance by the legend of the signature of Mohammed [68] which he traced with a tip of his sabre, or by the drawings reproduced by Listing [67] or Clausen [52].

> **Commentary.** Mohammed's signature (two crescents back-to-back) is drawn below, as can be seen as well in Lucas [2], *Récréations Mathématiques*, Vol. I (Paris, 1882).
>
>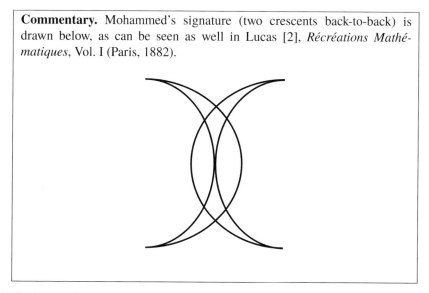

At the heart of such a study is the following obvious theorem: *the number of odd degree vertices of a graph is even* [68].

17. Euler has found that *any connected graph in which all vertices are even admits an Eulerian cycle, or a closed interlacing.*[52] *This is also the case if it only has two odd vertices* [25] [the case of an Eulerian chain, or *open interlacing*]. We must add,

---

[51] Such a graph is called an *entrelacement* (*interlacing*) by *S.-L.*

[52] The term *entrelacement*, used by *S.-L.*, could be inspired by the drawings of interweaving curves that cannot be separated, as we see in the article—
https://fr.wikipedia.org/wiki/Entrelacs_et_graphes

M. C. Golumbic, A. Sainte-Laguë, *The Zeroth Book of Graph Theory*, Lecture Notes in Mathematics 2261, https://doi.org/10.1007/978-3-030-61420-1_4

following Clausen [52] and Listing [67]: *if a graph has 2n odd vertices, we can find n chains, and no fewer, that use all the edges, using each one exactly once.* The proof of this proposition is easy [63, 68]. Moreover, Fleury [57] gave a practical procedure that allows us to find an interlacing in a given graph.[53]

Commentary. The Bridges of Königsberg

One of the best-known problems in graph theory is Euler's *Königsberg Bridges Problem* of 1736. Its solution is often regarded as the first result in graph theory. Much has been written about it.

For a full English translation of the original Latin version of Euler's paper, see, The Seven Bridges of Königsberg, in *The World of Mathematics*, (James R. Newman, ed.), Volume 1, pp. 573–580, or the book by Biggs, Lloyd and Wilson [230]. Recommended also are the two papers, Wilson [302] and Hopkins and Wilson [260].

A delightful student research report by Teo Paoletti (The College of New Jersey) entitled *Leonard Euler's Solution to the Königsberg Bridge Problem* can be found online in *Convergence*[a], a publication which offers a wealth of resources on teaching mathematics using its history.

---

[a] https://www.maa.org/press/periodicals/convergence/

---

[53] Fleury's algorithm [243] to compute Eulerian chains and cycles is still well known today. Interestingly, *S.-L.* cites Hierholzer's 1873 paper [60] in the Bibliography, but does not mention it in the text. It provides a different method for finding Eulerian chains and cycles. See also [242].

18. **Labyrinths**. — A *labyrinth* can obviously be represented by a graph. Trémaux [82] has shown that one can theoretically find a way out of a labyrinth—that is, follow all edges in a graph, whose structure is unknown, by applying a set of rules that he provided.

Maurice [70] showed that this can be done by using a different single rule.

Lucas [69] has noticed that the application of these rules allows us to transform such graphs into *trees* (§6). If, every time we arrive at a vertex A already visited, we step back and separate the edge leading to A from A, then the graph is transformed into a tree whose edges are all traversed twice. The tree depends on the chosen edges.

Commentary. Labyrinths, mazes and depth-first search

The reference to Lucas in this paragraph is about how we can find a way out of a labyrinth by exploring a graph using what we would call today a kind of randomized or greedy depth-first search (DFS). For a recent paper relating graph planarity to Trémaux trees (i.e., DFS trees), see [237].

At this point, we would have expected *S.-L.* to cite the paper by Tarry [79] which today is well known. He seems not to have known about an earlier paper by Christian Wiener [300]. An English translation can be found at the website by Michael Behrend from Somersham, Huntingdonshire, England that provides the re-publication of several historically important works of the mathematics of mazes (with English translation). These include the classic papers of Wiener (1873), Trémaux (1882), and Tarry (1895), as well as the chapter *The Labyrinth Problem* from Kőnig's book [268] that discusses in detail the maze-solving algorithms of Wiener, Trémaux, and Tarry. You can find his website at `https://www.cantab.net/users/michael.behrend/repubs/maze_maths/pages/index.html`

See also Volume 2 of Fleischner [242].

The reader interested in real mazes in gardens, cathedrals, archeological sites, amusement parks, and palaces throughout the world can visit the Labyrinth Society at `https://labyrinthsociety.org/` or the Labyrinth Resource Centre, Photo Library and Archive at `http://www.labyrinthos.net`, for a delightful virtual tour.

Commentary. From Édouard Lucas to George Lucas

The French mathematician, François Édouard Lucas, was interested in recreational mathematics [2, 8]—so much so, that he literally wrote volumes on it. Known mostly for his career in number theory, he studied and described the mathematics of many games, both simple and deep.

Lucas introduced the Tower of Hanoi puzzle, in his 1883 note, La tour d'Hanoï, under the pseudonym Professeur *N. Claus* (*de Siam*) Mandarin du Collège *Li-Sou-Stian*, an anagram of *Lucas d'Amiens* of *Saint-Louis*. An English translation by Paul Stockmeyer, and his extensive bibliography on the puzzle, can be found at—http://www.cs.wm.edu/~pkstoc/toh.html. A recent book on the topic is by A. M. Hinz, S. Klavžar, U. Milutinović and C. Petr, *The Tower of Hanoi – Myths and Maths* Birkhäuser (2013). See also, Paul K. Stockmeyer, The tower of Hanoi for humans, in *The Mathematics of Various Entertaining Subjects*, Volume 2: *Research in Games, Graphs, Counting, and Complexity*, Princeton University Press (2018), pp. 52–70.

The contemporary American filmmaker, George Walton Lucas, is best known as the creator of the Star Wars and Indiana Jones series of movies. However, as a student at University of Southern California in 1967, he wrote and directed a 15-minute science-fiction film, *Electronic Labyrinth: THX 1138 4EB*. The film won the 1968 United States National Student Film Festival drama award, and, in 2010, was selected for preservation in the United States National Film Registry by the Library of Congress. It was chosen for its enduring significance to American culture "with its technical inventiveness and cautionary view of a future filled with security cameras and omnipresent scrutiny." His film can be viewed at https://archive.org/details/electroniclabyrinth.

In 1986, George Lucas was the executive producer for Jim Henson's feature film *Labyrinth*. The main character, Sarah, must travel into a fantasy world to rescue her stepbrother from the Goblin King. Guarding his castle is a labyrinth, "a twisted maze of deception, populated with outrageous characters and unknown dangers". There is no known connection between Édouard Lucas and George Lucas aside from their interest in mazes.

19. Number of interlacings in a graph. — Delannoy [54], Tarry [80], and Lucas [68, 69] have calculated the number of distinct interlacings in a given graph.

Eulerian chains (§16) exist if and only if the number of odd vertices is 0 or 2. Supposing that there are $2n$ odd vertices, we reduce to this case by the introduction of a fictitious edge each time, when traversing a chain, we have to "jump" in order to access another one. Depending on the order in which we traverse the $2n$ edges, we find $N = \frac{(2n)!}{2^n n!} = 1 \cdot 3 \cdot 5 \cdot \ldots \cdot (2n-1)$ different combinations. Therefore, we can consider only graphs with even vertices, as we are going to do.

20. A graph with all vertices even can contain simple or multiple loops, that can be discarded first [80]. Let A be a vertex where $2q$ endpoints of q loops and $2p$ endpoints of other edges meet. We will see that if N is the number of Eulerian chains in the reduced graph,[54] then [the number of] Eulerian chains in the graph we started with is $p(p+1)\ldots(p+q-1)2^q$ times longer.

If A is a cut-vertex where a leaf block (§6) ends, and the leaf block has M Eulerian chains, which is necessarily an even number, and if $2p$ other edges end in A, we find that N is multiplied by pM. If there are q different leaf blocks that end in A, which have respectively $M_1, M_2, \ldots, M_q$ Eulerian chains, then the multiplication factor is $p(p+1)\ldots(p+q-1)M_1 M_2 \ldots M_q$. In this manner, a rose with n petals[55] has $2^n(n-1)!$ Eulerian chains, a graph formed by n circles pairwise tangent and with centers on the same line has 2^n, etc.

In a graph of strength 2 (§8), a subgraph is connected to a reduced subgraph by precisely two edges, one ending in A, the other in B. The multiplication factor that allows going from the reduced subgraph to the initial one is half the number of distinct Eulerian chains that are formed by removing the reduced subgraph and fusing together the two edges that had ended in A and B. If, for example, we connect in a quadrangle $ABCD$ the vertices A and B by $2p-1$ edges and C and D by $2q-1$ edges, then the number of Eulerian chains is $2(2p-1)!(2q-1)!$.

21. Consider next the method of decreasing the number of vertices by 1, due to Tarry [80]. Let A be a vertex of degree $2p$ that we want to remove. Joining pairwise in all possible ways the edges that end in this vertex, we obtain $N = \frac{(2p)!}{2^p p!}$ graphs that have fewer vertices than the original one.[56] In this manner, we will gradually reduce to the case of two vertices joined by $2n$ edges—that is, to the case where there are $2(2n-1)!$ Eulerian chains.

22. In this way, we can establish [69] that the number of Eulerian chains of a graph that consists of sides and diagonals of a regular polygon with $2n+1$ sides is the

---

[54] The term *"reseau reduit"*, used by S.-L. here, means the subgraph with loops removed, but is not defined in the manuscript.

[55] The term *"rosace à n feuilles"* probably refers to n cycles intersecting in one common vertex, but is not defined in the manuscript.

[56] *Corrected typo:* In the preceding formula, p incorrectly appeared as n in the original.

same as the number of Eulerian chains, up to n "jumps" (§19), for a polygon with $2n$ sides. Thus we find [61]:

pentagon $264 = 2^3 \cdot 3 \cdot 11$
7-gon $129\,976\,320 = 2^{11} \cdot 3 \cdot 5 \cdot 4231$
9-gon $911\,520\,057\,021\,235\,200 = 2^{16} \cdot 3^{11} \cdot 5^2 \cdot 7 \cdot 11 \cdot 40787$

The case of a 7-gon gives an answer to the following question, already solved in a more complicated way by Reiss [73]: *In how many ways can one lay out 28 dominos according to the rules of the game?* In a rectilinear layout of 28 dominos the last number is equal to the first one, and we can transform it into a circular layout which, after removing the doubles, gives a circular layout of 21 dominos. On the other hand, let us trace a regular 7-gon with vertices numbered $0, 1, \ldots, 6$ together with its diagonals; this gives a regular graph of degree 6 with even vertices. If we associate now, as proposed by Laisant [62], to each domino the corresponding diagonal or side of the 7-gon, to every circular layout corresponds an Eulerian chain of the graph, so we can compute the number sought:

$$2^{13} \cdot 3^8 \cdot 5 \cdot 6 \cdot 4231 = 2\,653\,076\,643\,840.$$

23. The figures formed by n equal circles pairwise tangent[57] have also been studied. If the centers of circles are vertices of a regular polygon with n sides, the number of Eulerian chains is $(n + 1)2^n$.

Consider also $2n$ circles of radius 1 with centers at the coordinates $(\pm 1, 2i)$, $i = 0, 1, \ldots, n - 1$.[58] Let $2^n U_{2n}$ be the number of Eulerian chains in the graph consisting of these circles. By adding to this graph a circle tangent to the first two circles, and denoting by $2^{n+1} U_{2n+1}$ the number of Eulerian chains in the new graph, Lucas [68, 69] gives the following relations:

$$U_{2n} = U_{2n-1} + U_{2n-2}, \qquad U_{2n-1} = 3U_{2n-2} + U_{2n-3}$$

If we write $U_{2n} = u_n$ and $U_{2n-1} = u_n - u_{n-1}$, these relations can be rewritten in the following reduced form:

$$u_n = 5u_{n-1} - u_{n-2},$$

which simplifies computation of the u_js, and, as a result, of the U_js.

Let us remark finally that, after Métrod [71], the number of Eulerian chains in a graph with 3 vertices connected pairwise by a, b, and c edges, respectively, these numbers being either all even or all odd, is

$$2b!c! \left(\frac{a+b}{2} - 1\right)! \left(\frac{a+c}{2} - 1\right)! \sum_k \frac{(a+k-1)!}{k! \left[\left(\frac{a-k}{2} - 1\right)!\right]^2 \left(\frac{b-k}{2}\right)! \left(\frac{c-k}{2}\right)!},$$

---

[57] *tangents deux à deux*, in the terminology of S.-L. We suppose he intends that two circles are either disjoint or tangent.

[58] *Corrected typo:* corrected from $i = 0, 1, \ldots, n$.

where k in the sum takes values of the same parity as a, b, and c, taking successively the values up to the smaller of the numbers b and c.

24. Complete cycles (Hamiltonian cycles) of a graph. — Several authors [74], and in particular Brunel [51], have studied the *complete cycles* (§5) of a graph. Brunel associates to each graph a *table* constructed as follows: Let a, b, c, d be vertices of a graph—for example, a tetrahedron. Let us name a the first row and first column of a 4×4 table, b the second row and column, and so on. For each edge, for example ab, put the corresponding letters in the corresponding cells of the table, a—b and b—a in our example (Fig. 3). If an edge ab is multiple (§4), we put $a_{b_1} + a_{b_2} + \ldots + a_{b_n}$ in the corresponding cell. In this setting, we observe that every *term* of such a table,

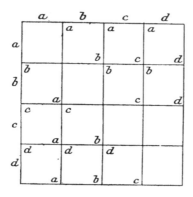

Fig. 3.[59]

in the usual sense of the word in the theory of determinants, represents precisely a complete cycle.

It is not always easy to find out how we can group the elements of the table to make terms out of them, or check whether it is even possible—that is, if the graph is *Hamiltonian* (§5).

25. Brunel considers two elements of a table, such as a_b and b_a, as being of opposite signs, and obtains a *left symmetric determinant*[60] if the number of vertices is $2n$. Let us call a *complete half-cycle* a set of edges, such as a_b, c_d, where every vertex appears precisely one time. If we take alternating edges of a Hamiltonian cycle, it decomposes into two complete half-cycles. The whole cycle is, in a sense, the product of the two *half-cycles*.

Theorems of Jacobi and Cayley on the left symmetric determinants allow us to write down an expression which gives a determinant when squared; this corresponds

---

[59] The entry in row c and column d is missing in the original manuscript. It should have entry c/d.

[60] *déterminant symétrique gauche*, in the terminology of S.-L.

precisely to looking at complete half-cycles. In this way, the previous case corresponds, up to sign, to the square of $a_b c_d + a_d b_c + a_c b_d$, a squared expression where a term such as $a_b c_d \times a_d b_c$ corresponds to a complete cycle a—b—c—d, a product of the two complete half-cycles a—b, c—d, and a—d, b—c.

Finally, the author shows that finding complete half-cycles can be reduced to finding complete half-cycles in a graph which has two fewer vertices.

26. In addition to this research, we can mention the work of Lemoine [63, 64] on the *communication problem*. In a graph, let us consider three groups of vertices: a group of n vertices, named A, a group of p vertices, named B, and a group of n vertices, named C, and consider n chains that connect vertices of group A to vertices of group C, with each point of A or C used precisely once. Each chain passes through $0, 1, \ldots, p$ vertices of B, in such a way that each vertex of B occurs in at most one chain. If $\gamma(n, p, q)$ is the number of chains that use q out of p vertices of B, then

$$\gamma(n, p + 1, q + 1) = \gamma(n, p, q + 1) + (n + q)\gamma(n, p, q),$$

from which it follows, observing that $\gamma(n, 0, 0) = n!$, that

$$\gamma(n, p, q) = n! \frac{p(p - 1) \ldots (p - q + 1)}{q!} n(n + 1) \ldots (n + q - 1).$$

The total number of chains is the total of the sum of the values $\gamma(n, p, q)$ where q ranges from 1 to p.

IV Regular graphs

27. Complete graphs. — The regular graphs that are the easiest to study are the *complete graphs* (§7) in which there exist $\frac{1}{2}n(n-1)$ edges that join all pairs of vertices. They have been studied by Sainte-Laguë [90]. The number of complete cycles is $\frac{1}{2}(n-1)!$. Their chromatic number (§7) is n. Their strength (§8) is $2(n-2)$ and their dimension (diameter and radius) (§8) is 1. We will see that their edge-chromatic number (§7) is $n = 2m + 1$ if n is odd, and $n - 1 = 2m - 1$ if n is even.

28. Polygonal graphs. — A polygonal graph is made of vertices of a regular polygon, numbered $0, 1, 2, \ldots, n - 1$, with edges being sides or diagonals of the polygon. The set of edges that join vertex 0 to vertex a, from 1 to $a + 1, \ldots$ form the *element* (a) of a graph (for $a \le n - a$).[61] If a is prime to n, it consists of a single polygon. A graph that has $(a), (b), \ldots$ for edges is connected only if $a, b, \ldots$ are pairwise coprime. *A polygonal graph is Hamiltonian* (§5, §57).

We find easily that if a graph of order n is of degree p, its edge-chromatic number is at least p if $n = 2m$ is even, and at least $p+1$ if $n = 2m+1$ is odd. In many situations, the edge chromatic number takes this minimum value. This is the case, for example, if $n = 2m + 1$, where $p = 2q$ is even, and if, moreover, the elements $(a), (b), \ldots$ are polygons. This is also the case if the greatest common divisors $\alpha, \beta, \ldots$ of a and n, b and n, $\ldots$, are pairwise coprime, and multiply up to n [90]. If $n = 2m$, the edge chromatic number takes its minimum possible value when, for example, the various elements are formed of polygons with an even number of sides, even if one of them contains the m diameters. The edge-chromatic number takes its minimum possible value if $n \le 14$. Perhaps this is so for all polygonal graphs!

29. The study of the chromatic number is one of the most complicated questions. We mention here several properties given by Sainte-Laguë [90]:

A polygonal graph of odd order and greater than 5, and consisting of two elements, is either tripartite or tetrapartite.

A polygonal graph of even order and consisting of two elements, is bipartite, tripartite, or tetrapartite.

---

[61] This means that $(a) = \{(i, i + a) \mid i = 0, 1, \ldots, n - 1\}$, where addition is modulo n and $1 \le a \le \lfloor n/2 \rfloor$.

© Springer Nature Switzerland AG 2021
M. C. Golumbic, A. Sainte-Laguë, *The Zeroth Book of Graph Theory*,
Lecture Notes in Mathematics 2261, https://doi.org/10.1007/978-3-030-61420-1_5

Finally, the necessary and sufficient condition for a polygonal graph of order n consisting of elements (a), (b), ... to be bipartite is that n be even and a, b, ... be odd.

30. The *strength* of a polygonal graph of order n and degree p can be less than $2(p-1)$ only when one can separate p vertices that form a maximal clique (§8). It is then equal to p, as is the case with the graph of order 10 and degree 5 that consists of three elements: (2), (4), (5).

The dimensions (§8) of a polygonal graph are the same for all vertices. There exist associable (§8) and even self-complementary polygonal graphs. The simplest is the graph of order 13 that consists of three elements: (1), (3), (4).

31. **Homeomorphic polygonal graphs**. — Let us trace a regular 7-gon and those diagonals that connect vertices at a distance of two sevenths of the circumference (Fig. 4). Number the vertices following the edges of the inscribed star polygon, and then draw a homeomorphic copy of the graph with vertices numbered in the natural way. We observe that these polygonal graphs look different, but are nevertheless essentially identical (§4). We say that there is a *regular correspondence* between these two graphs, but there also exist *irregular correspondences*.

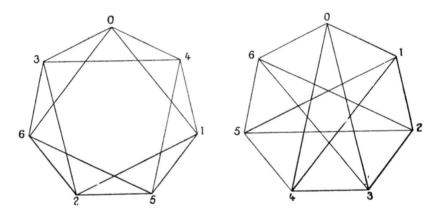

Fig. 4.

If we wish, as was done by Sainte-Laguë [90], to find the number of distinct polygonal graphs with n vertices, we have to study both regular and irregular correspondences. We will content ourselves with the first case here.

The first problem to solve is the following: *given a circle divided into equal parts, what is the number of distinct convex polygons with vertices among the division points?* Two such polygons, called (T) *polygons* are distinct if we cannot obtain one from another by a rotation of $\frac{2\pi}{m}$. The case of (T) polygons with one or two vertices is not excluded.

32. The second subject of research, treated in an equally preliminary fashion, is that of *semi-congruences*: two numbers a and b are called *semi-congruent* with respect to a modulus n, if either their sum or their difference is congruent to n, in the arithmetical sense of the word. This is denoted by

$$a \equiv b \pmod{n}.$$

We say that a number r *belongs to a semi-exponent q* if q is the smallest exponent for which r^q is semi-congruent to 1. In the notation of Gauss, such a minimum q is equal to m if m satisfies $2m = \varphi(n)$, the number of integers at most n and prime to it.[62] If $q = m$, then r is called a *semi-primitive root* of n.

Here are the principal results about semi-congruences:

- *The necessary and sufficient condition that every solution of one of the congruences or semi-congruences $x^2 \equiv 1$ and $x \equiv 1$ be a solution of the other is that their common modulus be semi-prime—that is to say, of the form a^α or $2a^\alpha$, where a is prime.*
- *The semi-exponent q to which r belongs, prime to n, is a divisor of m.*
- *If r belongs to a semi-exponent q, then this is also the case for r^ρ, if ρ is prime to q.*
- *There are 0 or $\varphi(n)$ semi-primitive roots of modulus n.*
- *All semi-primitive roots of n can be obtained from one of them, say r, by taking all numbers r^ρ for ρ less than m and prime to it.*
- *If r belongs to a semi-exponent q, on the one hand, and belongs to a semi-exponent q', on the other hand, then either $q = q'$ or $q' = 2q$.*
- *If n is semi-prime, then every primitive root of n is a semi-primitive root of n.*
- *A number n of the form 3^α admits a semi-primitive root 2, and $n = 2^\alpha$ admits a semi-primitive root 3. If a is an odd prime, then $n = a^\alpha$ admits 2 as a semi-primitive root ($\alpha > 1$), provided that $2^{\alpha-1} - 1$ is not divisible by a^2. Moreover, for all numbers less than 2000, only $a = 1093$ exhibits this divisibility [86, 87, 89].*
- *If r is a semi-primitive root of nn', then it is also a semi-primitive root of n.*
- *If $n = a^\alpha b^\beta c^\gamma \ldots$ has at least three distinct prime factors, then n has no semi-primitive root. This is also the case for $n = 2^\theta a^\alpha b^\beta$ with $\theta > 1$, or, if $\theta = 1$, provided that a and b are multiples of 4 plus 1.*
- *$n = 4a^\alpha$ or $n = 3a^\alpha$ admit semi-primitive roots, but $n = 8a^\alpha$ does not.*
- *If n is odd, then if one of the numbers n, $2n$ has semi-primitive roots, then the other has them too.*

33. To use the preceding results (§28), let us take a polygonal graph (R) of order n, that consists of p elements (a), (b), ... Let us consider the integers $\alpha, \beta, \ldots$ less than m, where $2m = \varphi(n)$, defined by

---

[62] The function $\phi(n)$ is often called *Euler's phi function*, or the *totient* of n. For $n \geq 3$, it is always an even number.

$$r^\alpha \equiv a, \ r^\beta \equiv b, \ \dots \quad (\text{mod } n),$$

and draw a circle divided into m equal parts, on which we consider points numbered $\alpha, \beta, \dots$ which are vertices of a polygon (T) (§31). If now we construct a graph (R'), homeomorphic to (R), by numbering the vertices from w to w where $w \equiv r^\omega$,[63] the new polygon (T') has points $\alpha + \omega, \beta + \omega, \dots$ as vertices, and can be obtained from (T) by a rotation of ω times an mth of the circumference.[64] The number of polygons (T) is the same as the number of graphs that do not become homeomorphic after renumbering from w to w.

The question is now settled if n has a semi-primitive root r, because each integer w is of the form r^ω. The question is more complex if n has no semi-primitive root, but it is still easy to compute the number $N = f(n)$ of distinct polygonal graphs. We have the following results:

$$n \ \ 4 \ 5 \ 6 \ 7 \ 8 \ 9 \ 10 \ 11 \ 12 \ 13 \ 14 \ 15 \ 16 \ 17 \ 18 \ 19$$
$$N \ \ 1 \ 1 \ 4 \ 2 \ 7 \ 5 \ 15 \ \ \ 6 \ \ 37 \ 13 \ 36 \ 32 \ 37 \ 34 \ 73 \ 58$$

$$n \ \ \ 20 \ \ \ 21 \ \ \ 22 \ \ \ 23 \ \ \ \ 24 \ \ \ \ 25 \ \ \ 26 \ \ \ 27 \ \ \ 28 \ \ \ \ 29 \ \ \ \ \ 30$$
$$N \ \ 183 \ 150 \ 262 \ 186 \ 1009 \ 420 \ 707 \ 703 \ 760 \ 1180 \ 4639$$

34. Primitive graphs. — Looking from a completely different point of view, Petersen [34] studied *primitive graphs*.

The study of invariants of binary forms led Hilbert [88] to consider products of the form

$$(x_1 - x_2)^\alpha (x_1 - x_3)^\beta (x_2 - x_3)^\gamma \dots (x_{n-1} - x_n)^\epsilon$$

where the degree of $x_1, x_2, \dots, x_n$ is the same for each variable, and some of the exponents can be zero. The primitive solutions of Diophantine equations correspond to the primitive factors of these products that are defined in an analogous way. Thus,

$$(x_1 - x_2)(x_2 - x_3)(x_3 - x_1)$$

is a primitive factor. Petersen represented each factor x_p by a vertex, and each term $x_p - x_q$ or $(x_p - x_q)^\alpha$ by 1 or α edges joining vertices x_p and x_q. The problem of finding the primitive factors is then formulated as follows: *Given a finite regular connected graph (R), with or without multiple edges, is it decomposable into regular graphs with the same vertices as (R)?* If this is not the case, then (R) is primitive, otherwise it is decomposable[65]. The graphs into which it can be decomposed are not necessarily connected.

35. Graphs of even degree. — Consider a graph of even degree $\alpha + \beta$ which can be decomposed in two different ways into graphs of degrees α and β. Let us take

---

[63] For example, Fig. 4 shows a numbering from 2 to 2.

[64] Namely, $2\pi\omega/m$.

[65] *Myriam Preissmann comments:* In other words, it (strictly) contains a spanning regular subgraph.

the first decomposition and color the edges of the first part *blue*, while the edges of the second are colored *red*. Passing from the first decomposition to the second, we recolor certain blue edges red, and vice versa. These red and blue edges form an *alternating cycle*, and all decompositions can be obtained in this way.

Let us take a graph (R) of order n and degree $2p$. We can always find an Eulerian cycle (§16) that has pn edges. If pn is even, then in coloring the edges of the cycle alternately blue and red, we get a decomposition of (R). This is why *every regular graph of degree 4 is decomposable*. If np is odd, so both n and p are odd, then there is no decomposition into graphs of degree p.

Let us now take two edges AB and CD in (R), with 4 distinct endpoints A, B, C, D, and replace them by AC and BD.[66] This gives a graph (R'), and we say that (R) and (R') are *twin graphs*. We can establish the following property: *if one of the twin graphs can be decomposed into graphs of degree 2, so can the other*. It follows that *any graph of even degree can be decomposed into graphs of degree 2*.[67]

36. In a graph (R) of order n and degree 4, consider two edges AB and CD. In general, we can choose a way of decomposing (R) such that AB and CD belong to the same component, or to different components of the decomposition. If this is impossible, then AB and CD are called *twin edges*. To determine whether AB and CD (Fig. 5) are twin edges or not, take an Eulerian cycle and represent it as a circle subdivided into $2n$ equal parts (Fig. 6), with segments colored alternately red and blue (in the figure, blue is represented by a double line, and red by a single line). Furthermore, let us join the *diagonals* by dotted lines, which are not edges of the graph, the vertices $AA, BB, CC, \dots$. They are *even* or *odd*, depending on the parity of subtended arcs.

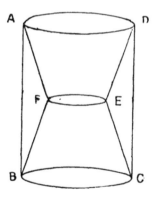

Fig. 5.

---

[66] S.-L. allows AC and BD to be multiple edges.

[67] *Myriam Preissmann comments:* In today's terminology: *the edges of a regular graph of even degree can be partitioned into edge-disjoint 2-factors.*

Finally, draw a line Δ, a *transversal line*, that meets the two edges AB and CD (Fig. 6). If the transversal line does not cut any diagonal, then AB and CD are twins. We also find that *if a transversal line intersects an even diagonal, AB and CD are not twins. They are twins if it intersects all odd diagonals and no even ones.* Petersen [34] found that this condition, which is necessary, is also sufficient.

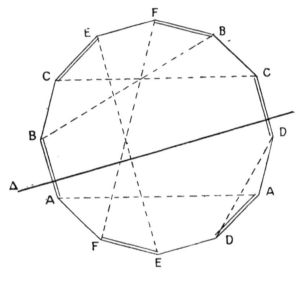

Fig. 6.

In the scheme that we have just constructed, erase any part that is not connected to the rest by two edges on the same side of Δ, obtaining a reduced scheme. We have the following property: *two twin edges AB and CD are of the same, or of different colors, depending on whether the number of vertices in the reduced scheme is even or odd.* Moreover, it is then possible to represent this scheme in such a way that there exists a line that intersects all edges, apart from AB and CD.

37. **Graphs of odd degree.** — There are no primitive graphs of even degree greater than 2 (§35), but on the other hand *there exist primitive graphs of any odd degree* $2p + 1$. For example, consider a triangle ABC with sides AB and AC, consisting of p [parallel] edges, and BC consisting of $p + 1$ [parallel] edges. Joining $2p + 1$ [68] similar triangles $ABC, A'B'C', \ldots$ by edges $AO, A'O, \ldots$ to a single point O, we construct a

---

[68] *Corrected typo:* The term $2p + 1$ appeared incorrectly as $2p$ in the original.
Myriam Preissmann comments: This is consistent with the order equal to $6p + 4$. The case $p = 1$ is indeed attributed to Sylvester by Petersen (see Biggs, Lloyd, and Wilson [230], p. 197.)

primitive graph of degree $2p + 1$ and order $6p + 4$. If $p = 1$, this graph is cubic and is sometimes called the *Sylvester graph*.[69]

Petersen [34] showed that *a regular graph (R) of order 2n and odd degree d greater than $\frac{2n+3}{3}$ is decomposable.*

Finally, Petersen [34] asked whether there is a simple way to determine in which cases a graph of odd degree is primitive. He believed that such a graph necessarily had leaf components (§6), but he limited himself to the case of cubic graphs, which allowed him to formulate the following property called the *Petersen Theorem*: *A cubic graph with fewer than three leaf components is reducible* (§42).[70]

We will find various other properties of regular graphs in (§40, §52).

Commentary. The Petersen graph

The Danish mathematician, Julius Petersen (1839-1910), constructed the following graph, now bearing his name—the *Petersen graph* [275]— illustrating the smallest cubic graph with no isthmus that has edge-chromatic number greater than three; see (§47) and its footnotes.

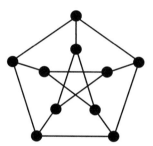

Over the years, the Petersen graph and its generalizations have served as useful examples and counterexamples for many problems in graph theory; see [232, 259, 301, 303].

The next chapter develops the topic of cubic graphs further.

---

[69] For a recent biography of Sylvester, see Parshall, K. H.: *James Joseph Sylvester: Jewish Mathematician in a Victorian World*, Johns Hopkins University Press, 2006.

[70] *S.-L.* never defined "reducible", but in this context the word seems to have the same meaning as "decomposable" which was defined at the end of (§34).

V Cubic graphs

38. **Bipartite cubic graphs**. — A particular case of regular graphs is that of cubic graphs (§7) which serves as a foundation for the statement of the *four color problem*. We will start with bipartite cubic[71] graphs—that is, those [cubic graphs] whose vertices can be partitioned into two sets so that two vertices of the same set are never joined by an edge. Such graphs are *bipartite* (§52) or of chromatic number 2 (§7). We assume that they do not have multiple edges.

The edge-chromatic number of a bipartite cubic graph is 3 (§53). Therefore, we can construct a bipartite cubic graph from even cycles by drawing diagonals in them, or by joining them by adding edges between pairs of vertices. If such a graph is of order $2n$, then it has $3n$ edges and its index (§7) is $n + 1$.

Sainte-Laguë [98] remarks, moreover, that a bipartite cubic graph is not always Hamiltonian.[72]

39. Given a bipartite cubic graph, denote by $A_0, A_1, \ldots$ the vertices of one of the sets and by $B_0, B_1, \ldots$ the vertices of the other. If A_0 is of *level* 0 (§8) then the vertices B adjacent to A_0 are of level 1, the vertices A, adjacent to these are of level 2, and so on. *The number N_k of edges that join a vertex of level $k - 1$ to a vertex of level k is a multiple of* 3. This follows from the identity

$$N_k + N_{k+1} = 3m_k,$$

where m_k is the number of vertices of level k (§8).

If $\alpha_k, \beta_k, \gamma_k$ are the numbers of vertices of level k with 3, 2, 1 incoming edges, respectively, from a vertex of level $k - 1$, then we have

$$3\alpha_k + 2\beta_k + \gamma_k = N_k, \qquad \alpha_k + \beta_k + \gamma_k = m_k, \qquad \beta_k + 2\gamma_k = N_{k+1}.$$

N_k is of the same parity as $\alpha_k + \gamma_k$ and lies between m_k and $3m_k$.

We can see that *a bipartite cubic graph cannot have an isthmus*, and that *if we partition the vertices and edges of a bipartite cubic graph into two groups that share no edges and only share r vertices from one and the same set A, then the total number of edges that join these r vertices to one or the other group of vertices is divisible by*

---

[71] *bicubique*, in the terminology of *S.-L.*

[72] For a more recent result, see Ellingham and Horton [240].

© Springer Nature Switzerland AG 2021

M. C. Golumbic, A. Sainte-Laguë, *The Zeroth Book of Graph Theory*,
Lecture Notes in Mathematics 2261, https://doi.org/10.1007/978-3-030-61420-1_6

3. All of these properties can be extended to regular bipartite graphs regardless of their degree.

40. **Hamiltonian bipartite cubic graphs**. — Sainte-Laguë [98] considers certain families of *Hamiltonian* bipartite cubic graphs (§5), which can be represented by a circle divided into parts by $2n$ points, named alternately A and B, and connected by n chords, each joining an A point with a B point.

If the n chords are the n diameters, then the graph is called *bidiametral*. Its dimension is $\frac{n+1}{2}$.

One can sometimes partition the n chords into two sets for which the chords from the same set do not intersect inside the circle, resulting in a *spherical* [planar] graph (§7). By grouping together the neighboring chords, we get a scheme consisting of a convex polygon cut arbitrarily by non-intersecting diagonals. In the case where each group encompasses a single side of the polygon, the graph is *semi-tiled bipartite*.

Assuming that $n = 2n'$, take a circle divided into $2n$ equal parts. If $n = v + w$, and $v = 2v'$, $w = 2w'$ ($v' \leq w'$), we may assume that the graph is formed by v parallel vertical chords and w parallel horizontal chords, for which each horizontal chord intersects each vertical one, and vice versa. Such a graph is called *tiled bipartite*, and the *dimension* (§8) of such graph can take any value between $v' + 2$ and $n' + 1$. If $v' = w'$, then the *diameter* of the graph is $2v' + 1$ and its *radius* is $v' + 2$. It is a balanced graph (§8) if and only if $v' = w' = 2$.

41. Bipartite cubic graphs with at most 16 vertices are all Hamiltonian [98]. We use the following notation to describe them. Denote by $A_0, B_0, A_1, B_1, \ldots$ the consecutive vertices of a cycle. The expression 5201 will denote a graph in which A_0 and B_5, A_1 and B_2, A_2 and B_0, etc., are joined by an edge.

In this notation, the only bipartite cubic graph with 6 vertices is 120; it is bidiametral.

There is only one bipartite cubic graph with 8 vertices: 2301;[73] it is semi-polygonal, tiled, and spherical [planar].

The two bipartite cubic graphs with 10 vertices are 23401 and 24301, and the first one is bidiametral. There are five bipartite cubic graphs with 12 vertices: 123450, 143052, 120453, 134520, and 140523; the first one is semi-polygonal. There are twelve bipartite cubic graphs with 14 vertices: 3456012, which is diametral; 2345601, which is semi-polygonal; 1250634, which is semi-tiled and spherical; 1205634, 3504162, 5430621, 3465012, 3465102, 5304621, 5241063, 5204163, and 2645130. Finally, here are 37 graphs with 16 vertices:[74]

---

[73] *Corrected typo:* The original had 2031.

[74] We have not verified the correctness of the longer sequences claimed in this section, but in the table below, 34067216 is certainly wrong because it has two 6s and no 5.

Myriam Preissmann comments: Since A_7 is already connected to B_6 by an edge of the Hamiltonian cycle, 34067216 should clearly be replaced by 34067215.

12345670 36402715 12305674 23456701 52741630
32017645 16042375 13075642 16542370 67042315
56742310 32047615 25317640 24317605 12406375
12407635 24361075 42360175 23641075 14306275
12605734 46572310 14562370 58716240 23516740
52701634 13546270 32547601 43052671 64052671
65402731 34067216 35716024 32457601 25460371
26501734 23496710

In the first line, the first graph and the two last graphs are semi-polygonal. The first one is also tiled and spherical, the second graph in the line is semi-tiled and spherical, and the third one is spherical.

42. **Petersen's Theorem**. — *A cubic graph with fewer than three leaf components is reducible* [34] (§37). The proof given by Petersen was significantly simplified by Brahana [93], and then by Errera [24, 95]. [75] We give here a proof analogous to the latter ones [99]; we will suppose, however, that there are no singularities: isthmuses or leaf components. This restriction is unimportant for applications to the *four color problem* (§1, §2). The theorem of Petersen can then be stated as follows: *The edges of a cubic graph with no isthmuses or multiple edges, can be partitioned into two colors in such a way that every vertex is an endpoint of one edge of the first color and of two edges of the second.* 43. Following Petersen, we call *red* all the edges of

the first color, or "R edges", and we call *blue* those of the second color, or "B edges". We will proceed by induction by assuming that the theorem is true for any graph with fewer vertices than the considered graph (ρ).

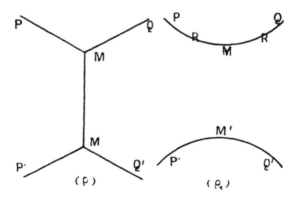

Fig. 7.

Consider the graph (ρ), and remove an arbitrary edge MM' from it, as well as the vertices M and M' (Fig. 7). This gives a graph (ρ_1), in which we partition the edges into R and B by the induction hypothesis. Two half-edges PM and MQ are naturally of the same color, as well as $P'M'$ and $M'Q'$.

Call an *alternating chain* a chain whose edges alternate in color. There are four possible cases: (1) PMQ and $P'M'Q'$ in (ρ_1) are B; in this case, it suffices to draw MM' and color it R. (2) There is an alternating chain that goes from M to M'; we then reverse the colors, and draw MM', colored B. (3) PMQ is R, $P'M'Q'$ is B, and there is an alternating chain that starts and ends at M. We then swap B and R and reduce to the case (1).

The only case left is (4), where at least one of the edges PMQ and $P'M'Q'$ is R; we assume without loss of generality that it is PMQ, and that no alternating chain starting at M ends in M or M'. We will show that this leads to a contradiction.

44. In Fig. 7, we have painted the edge PMQ in R. Let us now travel from M to either P or Q and call (ρ_2) the graph that contains those edges of (ρ_1) that belong to at least one of the alternating chains that start at M. All such chains are *oriented* (§5), with the *origin*, or *initial vertex*, being the first vertex reached after M, and the *ending*, or *final vertex*, the other vertex. The origin and the ending are colored R or B, depending on whether the considered edge is R or B. Certain edges, called *unicursal* (§5), whether they belong to one or several alternating chains, can only be traversed in one direction; others, called *bicursal*, can be traversed in both directions and have two distinct origins and two distinct endings.

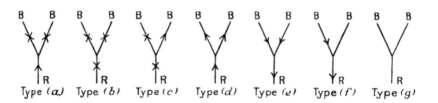

Fig. 8.

Thus, the different possible types of vertices $[(a), (b), \ldots, (g)]$ are those depicted in Fig. 8, where the direction of the arrow indicates the direction of the traversal of a unicursal edge, and bicursal edges are marked with a bidirectional arrow. The type (g), where none of the adjacent edges is included in an alternating chain, can occur in (ρ_1) but not in (ρ_2). There exist two linear relations between the numbers $a, b, \ldots$ of vertices of types $(a), (b), \ldots$ which can be found quite easily:

$$c - 2a + f + 4 = 0 \qquad \text{and} \qquad c + 2d - 2e - f = 0.$$

45. The value of $2a = c + f + 4$ cannot be zero, so there exist bicursal edges that form one or several connected graphs; let (ρ_3) be one of these graphs. It cannot contain any vertices other than $(a), (b)$ or (c). We cannot reach (ρ_3) other than via vertices (a), and we cannot leave it other than via vertices (c).

We find that (ρ_3), which contains at least one vertex of type (a), does not contain another vertex of type (a); the number of graphs (ρ_3) is therefore a.

In (ρ_3) all edges are bicursal, and each B edge ends in two distinct vertices, so the total number of B edges attached to the vertices of (ρ_3) must be even; therefore c is even and is at least 2, which implies $c \geq 2a$. This is a contradiction, because $c - 2a + f + 4$ is zero. This concludes the proof of the theorem of Petersen.

46. **Unicursal and bicursal graphs.** — We continue this line of research [99] and consider a cubic graph (ρ) in which, as above, two B edges and one R edge meet in every vertex. We see that *every vertex is incident to one of the alternating chains which, after leaving an arbitrary vertex M, starts with an R edge.* In the case that no chain returns there, we have to add a vertex of type (h). The equalities become

$$c - 2a + f + 2h = 0 \qquad \text{and} \qquad c + 2d - 2e - f = 0.$$

If (ρ') is contained (§5) in (ρ) and has only bicursal edges, then it has no more than one vertex (a) and at least two vertices (c).

This implies $c \geq 2a$, which implies in turn:

$$c = 2a, \ f = 0, \ h = 0, \ e = a + d.$$

Therefore, vertices of type (h) do not exist, and one of the alternating chains of M returns to this vertex. Moreover, because only one vertex (f) can connect the vertex (g) to other vertices, g is also zero. This finishes the proof of the proposition.

47. Let us limit ourselves to graphs of strength 4, which occur in the *four color problem* (§1, §2). A graph (ρ') does not exist, because it would not be connected to the graph (ρ), other than via a vertex (a) and two vertices (c), which makes three edges. Therefore, *in a cubic graph of strength 4* (§8), *all the edges are unicursal, or they are all bicursal, except for three edges that are incident to the same vertex.* In the first case the graph is unicursal, and in the second case it is bicursal.

We find [99] that in a unicursal graph, all alternating chains that connect two given vertices, have the numbers of edges of the same parity. All the cycles are even; therefore *every unicursal graph is bipartite*; this case is reduced to that of bipartite cubic graphs (§38). It is left to consider only bicursal graphs.

Let us also remark that an immediate consequence of the theorem of Petersen [99] is that a *cubic graph has edge chromatic number 3 or 4*. [76]

---

[76] *Myriam Preissmann comments:* In fact, it seems that Petersen's theorem implies directly only that a cubic graph with at most two leaf components has edge-chromatic number 3 or 4. Indeed, the more general version is true by Vizing's Theorem, but a proof may not have been known in 1926.

Finally, let us mention the *conjecture*[77] *of Tait* [100], which is closely related to the four color problem (§1, §2). It can be stated as follows: *A cubic graph with no an isthmus has edge-chromatic number* 3. The proof of this proposition is not known.[78]

---

[77] *S.-L.* calls it a *theorem*, but states that there is no known proof. However, see the next footnote.

[78] *Myriam Preissmann comments:* The Petersen graph illustrates that there are cubic graphs with no isthmus that have edge-chromatic number greater than 3—see Petersen [275] where he states (in a diplomatic manner) that there is indeed a counter-example to "Tait's theorem". *S.-L.* seems not to have known this yet in 1926; however, by 1929, he cited the paper in his work [285].

It is true that Tait believed that all cubic graphs with no isthmus have edge-chromatic number 3, but he proved only that, "Every *planar* cubic graph with no isthmus has edge-chromatic number 3 if and only if we can color the faces of every map with four colors." In fact, the only currently known proof of the result that "*every planar cubic graph with no isthmus has edge chromatic number 3*" relies on the proof of the four color theorem (Appel and Haken [225]).

VI Tableaux

48. Matrices. — The following idea can be found in the writings of many authors, including de Polignac [22], Brunel [51], Chuard [41], and Sainte-Laguë [122]. The idea is to associate with a given graph a rectangular table[79] (§10, §24). We will examine a representation proposed by Poincaré [117, 118, 119, 120] and studied by Veblen [125], Alexander [101] and Chuard [41], which we will analyze.

In a graph with *oriented* edges[80] (§5), a particular vertex may not belong to a particular edge, in which case the relationship between this vertex and this edge is represented by 0, or it can be an *initial vertex* of the edge, which is represented by +1, or a *terminal vertex*, which is represented by −1.

Let us associate to a graph (R) a table (T), where each of its p columns corresponds to an edge, and each of its n rows corresponds to a vertex. Each cell of the table is filled with 0, −1, or +1, depending on the relationship between the vertex and the edge. This table (T) is called the *incidence matrix*[81] of the graph (R).

49. Given the definitions above, we show that *the absolute value of the determinant of any square $r \times r$ submatrix Δ of (T) is 1 or 0*.[82]

To give a geometric interpretation to this study, we call the graph defined by Δ its *partial graph*. We say that a row of zeros represents an *isolated vertex* (§4); a column of zeros represents an *isolated edge*, with its endpoints removed;[83] a row with a single 1 or −1 gives a *pendant vertex* (§4); and a column with a single 1 or −1 gives a *pendant edge*, with one endpoint removed. It follows from what has been said that *the determinant of the incidence matrix of a partial graph that contains isolated vertices or edges is* 0.

---

[79] *S.-L.* uses the term *Tableaux* or Tables. In the early part of this chapter, it is used to mean the $\{1, 0, -1\}$ vertices-versus-edges incidence matrix of an *oriented* or *directed* graph. Later in the chapter, he uses the term differently for bipartite graphs.

[80] *S.-L.* uses the term *réseau orienté* (oriented graph) for the first and only time, referring to (§5) where he defined *chemin orienté* (oriented edge).

[81] *S.-L.* calls this *la matrice du réseau*, the "matrix of the graph", but we translate it into its common current term "incidence matrix" in order to avoid confusion with the "adjacency matrix" of a graph.

[82] *Myriam Preissmann comments:* In current terminology, this would be stated as *the incidence matrix of a directed graph is totally unimodular*. In this chapter, *S.-L.* sometimes writes "determinant" instead of "absolute value of the determinant".

[83] *amputé de ses extrémités*

© Springer Nature Switzerland AG 2021
M. C. Golumbic, A. Sainte-Laguë, *The Zeroth Book of Graph Theory*,
Lecture Notes in Mathematics 2261, https://doi.org/10.1007/978-3-030-61420-1_7

A *tree* (§6, §9) with a single vertex removed gives a determinant equal to 1. A polygon has an incidence matrix with zero determinant. If one adjoins various pendant edges [to the polygon], each adding one more vertex and one more edge, we get a *closed tree*,[84] which still has an incidence matrix of determinant 0. Finally, the value of the determinant of the incidence matrix of a graph does not change if one adjoins trees to it, or removes trees from it.

Let us call the *rank* of the matrix (T) of a graph (R) the number of rows or columns of the largest square submatrix with non-zero determinant.[85] *The rank of the incidence matrix of a [connected] graph of order n is equal to n − 1.* Recall that the *index* (§7, §15) is $p - n + 1 = \omega$.

50. **Linear equations and contours**. — Veblen [126], Alexander [101], and Chuard [41] have established a correspondence between systems of linear equations and graphs. Following Chuard [41], let us start with the incidence matrix (T) of a graph (R), and introduce as many variables $x_1, x_2, \ldots, x_p$ as there are columns. To each row of (T), and so to each vertex, corresponds an equation $\sum \epsilon_m x_m = 0$, where ϵ_m is $1, -1$, or 0, given by the column m. From another point of view, each edge is assigned an integer coefficient x, and we observe that around each vertex the sum of the coefficients is zero.

The system of n equations with p variables which we obtain in this way is the *system A* of the graph. Let us agree that the corresponding edge is traversed x times in the positive direction if x is positive, and $-x$ times in the negative direction if x is negative. Similarly, if an edge is not traversed at all, or if it is traversed k times in each direction, we call it a *null edge*.

A *contour*, and this term has a more general meaning than a cycle, corresponds to an algebraic sum $\sum x_m c_m$, where $c_1, c_2, \ldots, c_n$ denote the n edges. An algebraic sum of contours with coefficients is also a contour.[86] A *null contour* is a contour that contains only null edges.

51. The incidence matrix (T) contains an $(n-1) \times (n-1)$ submatrix with determinant equal to 1. We may assume that it corresponds to variables $x_1, x_2, \ldots, x_{n-1}$ and to the first $n - 1$ equations; this amounts to a tree in the graph. We have a set of solutions in x, described as follows:

Among the variables $x_n, x_{n+1}, \ldots, x_p$ [corresponding to the $p - n + 1$ edges that are not in the tree], set one of them, say x_r, to be 1, and all others to be 0. We solve the system and find that the values of $x_1, x_2, \ldots, x_{n-1}$ are always $-1, 1$, or 0. Depending on the choice of x_r, there are $p - n + 1 = \omega$ linearly independent solutions. The list of

---

[84] *arbre fermé*, in the terminology of S.-L., is used here for the first and only time.

[85] *rang de la matrice*, although S.-L. does not use the term again in the manuscript. Moreover, we believe the next sentence to hold only if the graph is connected.

[86] The term *"contour"* is not clearly defined. It seems to be a type of generalized cycle, being a linear or algebraic combination of cycles, or possibly simply what is sometimes called a circulation or algebraic flow. We might find a similar notion today in the theory of vector spaces or matroids, where they are simply called cycles. The term *"contour"* is used again later in (§62) with another meaning.

values of the xs (that are restricted to 1, -1, or 0) are solutions of the system A, and reciprocally, it would be easy to find here the properties of ω linearly independent cycles (§15).

The use of matrices and of systems of equations can be found in the study of *regions* [108, 109, 115, 116], or in the study of *schematic configurations* [114, 123, 124].

52. Semi-regular matrices (*Tissus*).

— The case of a bipartite graph (§7) is considered in Sainte-Laguë [122]—namely, the set of vertices is partitioned into two sets A and B, so that a vertex of A can be connected only to a vertex of B, and vice versa. If there are m vertices in A, and n vertices in B, we can construct a table of m rows, one for each vertex in A, and n columns, one for each vertex in B, with 1 in each cell that corresponds to an edge that connects a vertex of A to a vertex of B, and 0 (or sometimes -1, depending on convention) otherwise. Interchanging 0 and 1 changes the graph into its *associated graph*[87](§8).

A *semi-regular matrix*[88] is a table in which the number p of 1s in each row is constant, and also the number q of 1s in each column is constant ($mp = nq$); this corresponds to a *bipartite semi-regular graph*. If $m = n$, and so $p = q$, we have a square semi-regular matrix that represents a bipartite regular graph—or, for short, a *biregular graph*.

53. Using semi-regular matrices we can prove that *the edge-chromatic number of a biregular graph is equal to its degree*, or more generally, *the edge-chromatic number of a bipartite graph is equal to the maximum of the degrees*. We will point out later (§55) a proof due to Dènes Kőnig [106], different from the one that used semi-regular matrices.

This proposition is essentially equivalent to saying that in the determinant of a matrix in which the number of 1s is constant in each column or row, one can always find a non-zero term.[89] Removing this subgraph of degree 1 with n edges, we replace the initial biregular graph by another one, from which we then remove again a subgraph of degree 1, and so on. From this follows a proof that we will not reproduce here [122].

---

[87] In modern teminology, its *bipartite complement*.

[88] *tissu* is the term used by *S.-L.*—namely, a binary matrix with constant row and column sums. In French, the word *tissu* means fabric or cloth, the matrix reminding one of woven textiles. Lucas introduced the topic of *géométrie du tissage* in his papers [111, 112, 113] which *S.-L.* listed in the bibliography. The idea was indeed inspired by principles of weaving fabric with rectilinear threads.

The *geometry of fabrics* has become a well-studied area, branching off into many mathematical directions of research. For a good introduction, the reader is referred to the papers of Branko Grünbaum and Geoffrey C. Shephard [257, 258] and the extensive 1999 bibliography by Joseph Malkevitch at https://www.york.cuny.edu/~malk/biblio/geo-fabrics-biblio.html. See also [226, 238] and the *Commentary* introducing Chapter VIII. The term "semi-regular matrix" was introduced by Brualdi [233]. See also Gropp [253].

[89] *Myriam Preissmann comments:* Equivalently, there exists a perfect matching in the biregular graph.

54. In order to study semi-regular matrices, it is convenient to fix the number of rows. The semi-regular matrix is then of *width m*, and of *length n*. In a *mi-partite*[90] *semi-regular matrix*, *m* and *n* are even: $m = 2q$, $n = 2p$. There are q (+1)s and q (-1)s per column, p (+1)s and p (-1)s per row.[91] The number of different columns that can be formed is

$$2Q = C_{2q}^{q} = \frac{(2q)!}{(q!)^2}.$$

These [columns] are pairwise *associated* [complementary].

By definition, an *elementary semi-regular matrix of width m*, cannot be decomposed into semi-regular matrices of the same width. Two associated columns A_k and A_k' form an elementary semi-regular matrix. Consider, for example, semi-regular matrices of width 8, for which $Q = 35$.

Denote by $X_1, X_2, \ldots, X_7, A_1, A_2, \ldots, A_{28}$ a set of 35 out of 70 columns, and by $X_1', X_2', \ldots, X_7', A_1', A_2', \ldots, A_{28}'$ their associated columns. The choice of the first 35 columns is arbitrary, and the choice of the columns in X can be made in several ways, but subject to certain restrictions.

Denote by x_i or a_i the number of times we use a particular column X_i or A_i in a semi-regular matrix, if x_i or a_i are positive; and X_i' or A_i' if they are negative.[92] Writing down the condition that we indeed have a semi-regular matrix, we obtain 8 equations, from which one can be removed because it follows from the others: $\sum x_i + \sum a_i = 0$.[93] Although using a completely different interpretation, we have reduced to a computation analogous to those of the system A (§50). Taking the x as variables, and taking into account that the determinant is 1, if the columns X are well chosen, we can express the xs as functions of the as with integral coefficients. We thus obtain a result that can be formulated as follows: *There exists a non-canonical system of basic semi-regular matrices that, if juxtaposed, give all possible semi-regular matrices, after adjoining algebraically*[94] *an arbitrary number of pairs of associated columns. A given mi-partite semi-regular matrix can be obtained in this fashion in a unique way.* Consideration of such basic semi-regular matrices, of which there are 28 in the example above, shows that elementary semi-regular matrices can be of arbitrary length; but if no column can occur more than once, this is no longer the case, and a semi-regular matrix of width 6 cannot have length other than 2, 4, 6, or 10.

---

[90] *mi-partite*, in the terminology of S.-L.

[91] Notice that here S.-L. chose to use a $(-1, +1)$-matrix and not a binary matrix.

[92] *Corrected typo:* We replaced "if X_i or a_i are positive" in the original manuscript by "if x_i or a_i are positive", since it is clear that S.-L. meant the number x_i and not the column X_i. Similarly, in the summation in the next sentence, we replaced X_i by x_i. However, see the commentaries by Myriam Preissmann in the next footnote and on the following page.

[93] *Myriam Preissmann comments:* It seems to me that the equation $\sum x_i + \sum a_i = 0$ is not correct either. My interpretation is the following: the 8 equations are $\Sigma \epsilon_{i,j} x_i + \Sigma \alpha_{i',j} a_{i'} = 0$ where $\epsilon_{i,j}$ is the value of the j-th coordinate of X_i and $\alpha_{i',j}$ is the value of the j-th coordinate of A_i for $1 \leq j \leq 8$, $1 \leq i \leq 7$ and $1 \leq i' \leq 28$ (similarly to what is done in (§50)). The fact that each column vector has the same number of +1 and −1 entries implies that if there exist x_is and a_is such that 7 of the equations are satisfied, then the remaining equation is indeed satisfied too.

[94] The term *adjonction algébrique* (adjoining algebraically) was left undefined.

In the case of a non-mi-partite semi-regular matrix, we can reduce to similar computations and have equations with coefficients 1 or 0, with determinants formed from 1s and 0s.

Semi-regular matrices have been studied from a somewhat different point of view by Lucas [111, 112, 113] and Gand [104].

Commentary.

Myriam Preissmann comments: Below is my understanding of what I think Sainte-Laguë wanted to say.

Sainte-Laguë considers here bipartite graphs for which one part, (say M), has $m = 2q$ vertices of degree p, where m is fixed, and the other part, (say N), has $n = 2p$ vertices of degree q. Let us call such a graph a *tissu* of width m. It is represented by an $m \times n$ matrix whose entries are either $+1$ or -1. If the vertex m_i is adjacent to the vertex n_j, then the ij-entry is $+1$; otherwise, it is -1. Each column contains q $(+1)$s and q (-1)s and each row contains $p = n/2$ $(+1)$s and $p = n/2$ (-1)s. Adding or removing two associated columns (that is, two vertices of degree q for which the union of their neighbors is M) does not change the property of being a tissu of width m. Notice that multiplying a column by -1, we obtain its associated column. So an elementary semi-regular matrix (of length greater than 2) contains no associated columns.

Hence, it seems to me, $X_1, \ldots, X_7, A_1, \ldots, A_{28}$ are chosen in such a way that no two of these columns are associated. So, Sainte-Laguë should have written something like, "we choose arbitrarily 35 pairwise non-associated columns". Given a semi-regular matrix which does not contain two associated columns, if the X_i-column (resp., A_i-column) appears a positive number of times in the matrix, then x_i (resp., a_i) is set equal to the number of times the column appears; if the X_i'-column (resp., A_i'-column) appears a positive number of times in the matrix, then x_i (resp., a_i) is set equal to minus the number of times the column appears; and if none of X_i and X_i' (resp., none of A_i and A_i') appears in the matrix, then x_i (resp. a_i) is set to 0.

Finally, Sainte-Laguë remarks that if the 7 columns in X (or, more generally, $m - 1$ columns in X) are linearly independent columns, then we are able to deduce the x_is from a knowledge of the a_is.

55. **Bipartite graphs**. — One can also study bipartite graphs directly. Dènes Kőnig [106] has shown (§53) that *the edge-chromatic number of a bipartite graph is equal to the maximum of the degrees of the elements of the partition*, using a classical property mentioned by many researchers [41, 106, 122]: in a bipartite graph all cycles are even.

The author then derived the following properties of number tables: *if all elements in a determinant are positive integers or zero, and in every column or row the sum of elements is constant and non-zero, then there is at least one term in the determinant which is not zero*, a property mentioned above (§53, §54). Let us remark finally that the same author has considered *infinite* bipartite graphs (§7).

56. The following proposition is due to Errera [55]: *If we do not allow the edges to intersect each other, the maximum number of simple edges that connect n points of a set A in the plane with m points of a set B is* $2(n + m - 2)$.

Finally, Hadamard [28] has asked the following question: *What is the number of bipartite graphs of order 2n and degree 2 ?* Considering the number r of distinct cycles that cover the graph, he finds the formula:

$$N_{n,r} = N_{n-1,r-1} + N_{n-r,r}$$

where $N_{n,r}$ is the number of graphs with r cycles. Using this formula, he finds the following values $N = f(n) = \sum_1^n N_{n,r}$ for small n:

n	1	2	3	4	5	6	7	8	9	10
N	1	2	3	5	7	11	15	22	30	42

n	11	12	13	14	15	16	17	18	19	20
N	56	77	101	135	176	231	297	385	490	628

VII Hamiltonian graphs

57. Hamiltonian graphs. — Hamiltonian graphs[95] are graphs in which there exists a complete simple cycle (§5). Sainte-Laguë [148] has found that *every polygonal graph is Hamiltonian*. This is obvious if at least one of the *elements* (§28) of the graph is formed from a single polygon. Otherwise, we take two coprime elements $(a), (b)$ that are not coprime to n, and give the partial graph a *toroidal* representation (§7), where (a) are meridians and (b) are parallels.

From this we derive a tiling which helps to prove the statement.

For small values of n, bipartite cubic graphs are Hamiltonian (§41), but this is not true for any n; moreover, *every graph with an isthmus is not Hamiltonian.*[96]

Moreover, if we examine the simplest regular graphs [148], we find that for degree 2 they are all Hamiltonian. The degree 3 graphs are Hamiltonian graphs for $n = 4, 6, 8$; for $n = 10$, out of 19 graphs, two are not Hamiltonian, of which one has an isthmus; for $n = 12$, out of 80 graphs, five are not Hamiltonian, of which four have an isthmus. For degree 4 [149], $n = 5, 6, 7, 8, 9$ give 16 graphs, all Hamiltonian, and $n = 10$ gives 57, of which two are not Hamiltonian.

58. Permutations. — Permutations have been extensively studied, and we refer in particular to the interesting overview of Aubry [127]. This area of research immediately leads to the study of graphs. Let us number $1, 2, \ldots, n$ the consecutive points that divide a circle into n equal parts. To each permutation $pqr \ldots$ of these n numbers, we associate the polygon with vertices $pqr \ldots$ which we call the *polygon of the permutation*. By adding a circle to it, we get the *graph of the permutation*, a regular Hamiltonian graph, of order n. The reciprocal statement would not be true, because such a network could be of *strength* 2 (§8).

Sainte-Laguë [150] classifies permutations according to the properties of the graphs corresponding to them. He simultaneously considers the representation of Lucas [142] who considers a square table with n lines and n columns, and associates to a permutation $pqr \ldots$ the p-th cell of the first column, the q-th cell of the second column, etc.

---

[95] *Réseaux cerclés*, in the terminology of *S.-L.*

[96] Recall that in (§39) *S.-L.* noticed that a bipartite cubic graph cannot have an isthmus.

© Springer Nature Switzerland AG 2021
M. C. Golumbic, A. Sainte-Laguë, *The Zeroth Book of Graph Theory*,
Lecture Notes in Mathematics 2261, https://doi.org/10.1007/978-3-030-61420-1_8

To simplify the exposition, we consider a particular example to illustrate this parallel study of objects of different kinds, associated with a permutation A: 1245637, its *graph* (R) or its *polygon* (P) (Fig. 9), and its *table* (T) (Fig. 10).

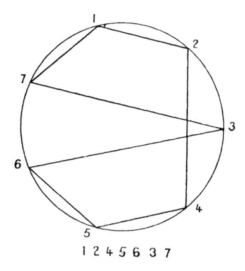

Fig. 9. The permutation A

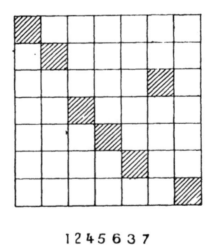

1 2 4 5 6 3 7

Fig. 10. The table for the permutation A

59. The *circular permutations* of A: 1245637, namely, 2456371, 4563712, etc., give the same graph (R); the new tables can be constructed from (T) by [circularly] permuting the columns.

The *additive permutations* of A, namely, 2356741, then 3467152, etc., where each consecutive instance is derived from A by applying a circular permutation to each element in the presentation of A,[97] [similarly] give the same graph (R). The corresponding tables can be obtained from (T) by [circularly permuting] the rows.

The *inverse permutation* of A, namely, 7365421 also gives the same graph (R) and table (T), by symmetry with respect to a vertical axis.

Finally, the *complementary permutation* of A, namely, 7643251, where we subtract each digit from 8 in the representation of A, gives again the same graph (R) and table (T), by symmetry around a horizontal axis.

To each permutation A of n elements, there corresponds $4n^2$ permutations of the group (a) which are not always distinct, as one observes with 1234567. They all have the same graph (R), and the tables can be derived easily from the table (T). Moreover, it is better to consider here, not (T), but rather (θ), the *infinite table*, formed by juxtaposition of copies of (T) and thereby covering the whole plane.[98]

60. We now introduce the notion of *reciprocal permutations* [127, 150]. Let us write the reciprocal permutation A' below A:

$$A: \quad 1 \ 2 \ 4 \ 5 \ 6 \ 3 \ 7$$
$$A': \quad 1 \ 2 \ 6 \ 3 \ 4 \ 5 \ 7$$

In A', the ordering $1, 2, 6, 3, 4, 5, 7$ indicates the positions in A occupied by the integers, consecutively.[99] We will see that A' can be equal to A, [but generally is not].

The reciprocal permutation of A' is A itself. The reciprocal permutation of the permutation complementary to A is the inverse permutation of A'. The permutations reciprocal to circular permutations of A are the additive permutations of A'. To a group (a) there corresponds, by reciprocity, a group (a'). The union of these groups, which consists of at most $8n^2$ distinct permutations, gives the *permutations of a group* (A).

To obtain the reciprocal graph (R') from a graph (R), we number the vertices of the polygon (P): 1234567, and then draw the second homeomorphic graph, in which (P) becomes the exterior circle: this is the desired (R'). So, a single graph and some of the graphs homeomorphic to it are sufficient to represent all $8n^2$ permutations of the group (A). We deduce, for example, that because the number of double edges is the same in (R) and (R'), *the number of sides of polygons subtending [inscribed into] a single arc of the circle is the same for two reciprocal graphs.*

---

[97] that is, by adding 1 to each number in the previous instance

[98] *S.-L.* uses the term *tableau indéfini* (indefinite table).

[99] For example, the number 3 appears in position 6 of A, so the number 6 appears in position 3 in A'. The table for A' is obtained by reflecting the table for A through the main diagonal.

The reciprocal tables (T) and (T'), or (θ) and (θ'), are symmetric with respect to a main diagonal.

A is a *self-reciprocal permutation* if it is identical to A'. In this case, (T) is identical to its symmetric reflection (§61, §71).

The classification of the initial permutations, if we do not take into account the reciprocity, gives the following table:

n	$n!$	groups of $2n$	groups of $4n$	groups of n^2	groups of $2n^2$	groups of $4n^2$
3	6	1 of 6				
4	24	1 of 8	1 of 16			
5	120	2 of 10			2 of 50	
6	720	1 of 12	1 of 24	3 of 36	6 of 72	1 of 144
7	5040	3 of 14			21 of 98	15 of 196
8	40320	2 of 16	1 of 32	11 of 64	67 of 128	121 of 256

If we take reciprocity into account, then this table becomes:[100]

n	$n!$	groups of $2n$	groups of $4n$	groups of n^2	groups of $2n^2$	groups of $4n^2$	groups of $8n^2$
3	6	1 of 6					
4	24	1 of 8	1 of 16				
5	120	2 of 10			2 of 50		
6	720	1 of 12	1 of 24	3 of 36	2 of 72	3 of 144	
7	5040	1 of 14	1 of 28		9 of 98	19 of 196	8 of 392
8	40320	2 of 16	1 of 32	7 of 64	21 of 128	47 of 256	49 of 512

61. Directed cycles. — The notion of a permutation is closely associated with the classical notion of a *directed cycle*, due to Cauchy [134], which results from the comparison of a permutation with a monotone sequence of integers. Thus, A: 1245637 contains three directed cycles of length 1: 1—1, 2—2, 7—7, and a directed cycle of length 4: 4—3—6—5. As was shown by Sainte-Laguë [150], the notion of directed cycles is related to the schemes introduced previously. For example, if we draw in the same table the permutation A and the permutation 1234567, then

---

[100] *Myriam Preissmann comments:* S.-L. considers the equivalence classes of the permutations depending on the value of n (n is indicated in the first column, the second column is the total number of permutations). In the first table, he considers that two permutations are equivalent if they can be obtained one from the other by using circular permutations, additive permutations, inverse permutations, complementary permutations (all groups (a), each has at most $4n^2$ members). In the second table, he adds reciprocal permutations (all groups (A), each has at most $8n^2$ members).

we will have (Fig. 11) the three directed cycles of length 1, denoted here by points with a circle, and a directed cycle of length 4 which gives the polygon with 8 sides, drawn here with a dotted line.

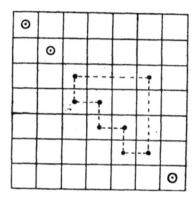

Fig. 11.

We can check that: A, *its inverse, its complement, and its reciprocal have the same directed cycles.*[101] *A permutation is self-reciprocal if its does not admit directed cycles other than those of length 1 or 2.*

To a schema such as that of Fig. 11, we can associate a new graph (ρ) by drawing a circle divided into 7 equal parts. The directed cycle 1—1 gives the vertex 1 with a loop, the same is true for 2—2 and 7—7, and the directed cycle 4—3—6—5 gives a rectangle with vertices 4, 3, 6, 5. These schemas are almost identical to those that appear in the following problem.

62. **The problem of postage stamps.** — In Lucas [142] we find the following question: *In how many ways can one fold a strip of postage stamps?* Despite its apparent simplicity, this question remains unsolved.

If we decide, before folding, to number the stamps $1, 2, \ldots n$, then the folded strip will give, from top to bottom, a certain permutation. We have to distinguish *permutations that can arise this way from those that cannot.*

---

[101] This is false: the directed cycles are *not* the same for all four permutations. Rather, A and its reciprocal A' have the same cycles, and the inverse of A and the complement of A have the same cycles.

Commentary. A postage stamp strip from the non-existent, fictitious country of Timbrelande. Original artwork by the younger author.

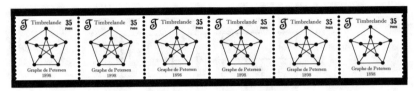

French postage stamps that André Sainte-Laguë surely must have seen while writing *Les Réseaux* (*ou Graphes*) are the set of four stamps *Les Jeux Olympiques d'été de 1924*, commemorating the 1924 Summer Olympics in Paris. The stamps were designed by the artist Edmond Henri Becker and can be viewed at the website `https://www.phil-ouest.com/Timbres.php?Cas=Artiste&ListeMots=Becker&Ordre=DateVente` or in the personal collection of the younger author.

The 1924 Olympics were the first to use the official marathon distance of 42.195 km (26.219 miles), fixed by the International Amateur Athletic Federation (IAAF) in May 1921, and the standard 50m pool with marked lanes. During the games, British runners Harold Abrahams and Eric Liddell won the 100m and 400m events, respectively. Their stories are depicted in the 1981 movie *Chariots of Fire* whose title is said to be inspired by the line, "Bring me my chariot of fire!", from a William Blake poem, and the original Biblical phrase רֶכֶב אֵשׁ in II Kings 6:17.

As Sainte-Laguë proposes [151], let us represent each stamp of the strip by a point on a line, and represent by a half-circle (or a topologically analogous contour) the fold that connects each stamp to the following one. In this way, we associate with a possible permutation, such as

$$2—1—3—6—5—4—7—9—11—12—10—8$$

a schema such as the one in Fig. 12. In this way, we obtain a curve which, if the

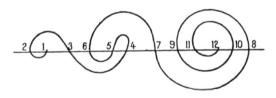

Fig. 12.

half-circles are placed alternately on the opposite sides of the line, has no double points. The opposite statement is obvious.

We limit ourselves to the following remarks: *if a permutation A is possible, then any circular permutation of A is possible.* This theorem is fundamental, because it highlights the only simple group that consists of such permutations. It allows us to assume that every permutation starts with 1. *If A is possible, then the inverse permutation and the complementary permutation of A are also possible.* To simplify what follows, let us number the points of the line (Fig. 12) in the usual order, and then connect them in the order in which they are met by the curve. [In this way,] we obtain new permutations, such as the following one [for Fig. 12]:

$$2—1—3—6—5—4—7—12—8—11—9—10$$

and the following statements:

The number K_n^p of possible permutations that start with $(1, p)$ is equal to the number that start with $(1, n + 2 - p)$. This can be concisely written down, with the obvious notation, as

$$K_n^p = K_n^{n+2-p}.$$

The number of possible permutations that start with $1, 2$ is equal to the total number that start with 1 and which have one fewer element. More precisely,

$$K_n^2 = K_n^n = K_{n-1},$$

where

$$K_n = K_n^2 + K_n^3 + \cdots + K_n^n.$$

If n is even, then there is no possible permutation starting with 1 and followed by an odd number. This can be written as:

$$K_{2p}^{2q+1} = 0$$

It is easy to find families of possible permutations—for example, by taking an arbitrary sequence of increasing integers, and then, after arriving at n, taking all the remaining integers in decreasing order. We can also construct a possible permutation, in many different ways, starting from another. If, for example, a permutation ends with

$$\ldots, n - 2q, n - 2q + 2, \ldots,$$
$$n - 2, n, n - 1, n - 3, \ldots, n - 2q + 1.$$

we can replace these terms by[102]

$$\ldots, n - 2q, n - 2q + 1, n - 2q + 3, \ldots,$$
$$n - 3, n - 1, n, n - 2, n - 4, \ldots, n - 2q + 2.$$

If between p and $p + 1$, there are integers $r, s, \ldots, u$ for which the list

---

[102] *Corrected typo:* The term $n - 3$ appeared incorrectly as $n + 3$ in the original.

$$\ldots, p, r, s, \ldots, u, p + 1, q + 1, \ldots$$

contains, between r and u, all the integers from $p + 2$ to q, then we can replace this list with

$$\ldots, p, p + 1, u, \ldots, s, r, q + 1, \ldots.$$

We also remark that *if a permutation is possible, then it cannot be produced from the problem of n pairwise non-attacking queens on a chessboard of n^2 cells* (§72).

63. We can give various recurrence formulas. For example, let I_n be the number of impossible permutations, and ω_p be the number of orderings of p stamps, where the impossibility arises only from the last stamp. We will establish the following formulas:

$$I_n = 5, 6, \ldots (n - 1)\omega_4 + 6 \cdot 7 \ldots (n - 1)\omega_5 + \cdots + (n - 1)\omega_{n-2} + \omega_{n-1},$$

$$I_n - nI_{n-1} = \omega_n.$$

The first values of $N = f(n)$, the number of possible permutations of n stamps, are given in the following table:

n	1	2	3	4	5	6	7	8	9	10
N	1	2	6	16	50	144	448	7472	17676	41600

Finally, consider the upper half of the schema in Fig. 12—namely, taking only the upper halves of the circles. After closing the line with itself, we are brought back to consider a circle and chords that do not intersect. In this way, we encounter, up to several restrictions, a bipartite cubic graph (§40), in a form that we often find in various topological problems.

64. **Curves**. — The preceding questions are related to various topological questions. Many of them concern the figures that are formed by a thread put on a plane, without knots, and with only crossing points[103] [127, 128]. We have, for example, the following statement [152]: *On a curve with no break points and only crossing points, we can mark at every crossing point a plus sign + on one of the two intersecting lines and a minus sign − on the other, in such a way that if we follow the curve, then we encounter + and − alternately.*

Let us also cite the theory of *characteristics* due to Kronecker [139], which we summarize following Weber [153].

For each closed curve without crossing points,[104] we call the *positive direction* (for example) the direction for which the interior of the curve is on the left side. Let us take two curves φ and ψ and call an *enclosure* the part of the plane interior to one curve and exterior to the other one.[105] An *entry point* is a point through which the

---

[103] with at most one crossing at any point in the plane, called *points doubles* by S.-L.

[104] that is, with no crossings.

[105] that is, the symmetric difference of their enclosed areas.

curve passes, in the positive direction of φ, entering the interior of ψ. An *exit point* can be defined in the same way. Now add a third curve f (Fig. 13). Let us associate with the positive direction on f, φ, and ψ an oriented cycle $f, \varphi, \psi, f, \varphi, \psi, f, \ldots$ (or, in short, f, φ, ψ), and to the negative direction the inverse oriented cycle f, ψ, φ. For

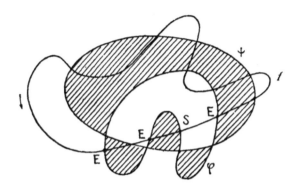

Fig. 13.

the oriented cycle f, φ, ψ, we define E to be an entry point of f in the enclosure φ, ψ, [the shaded area in Fig. 13] provided that it lies on φ and not on ψ. Similarly, we define S to be an exit point of f in the enclosure φ, ψ, provided that it lies on φ and not on ψ. In Fig. 13, there are three entry points E and one exit point S.[106] The *characteristic* k of the system of curves f, φ, ψ[107] is one half of the difference $|E| - |S|$. Here, $k = 1$.

If we denote this number by $k(f, \varphi, \psi)$, then we establish that there is only one value that k can take for an oriented cycle f, φ, ψ.

65. Braids. — Consider n straight lines, $1, 2, \ldots, n$ drawn on a plane, such that no two of them are parallel and no three of them intersect at a single point[108] [150]. Let us cut them by a moving *transversal* line Δ that sweeps across in a manner parallel to itself, giving rise to various permutations. Each time that it passes through a point common to two lines, we check that the number of *inversions* changes by 1. This allows us to study the properties of inversions of permutations [127].

We can easily generalize this by considering a *braid*[109] [150], formed not by straight lines, but by arbitrary curves, cut by a transversal line or by a curve similar to it from a topological point of view.

---

[106] *Myriam Preissmann comments:* However, looking at the labeled crossing points in Fig. 13, there seems to be another (unlabeled, second) exit point S, where f leaves the enclosure φ, ψ, the shaded area, through a point in φ, followed immediately by another (unlabeled, fourth) entry point E, where f crosses φ passing again into the enclosure φ, ψ.

[107] *Corrected typo:* Corrected from f, φ.

[108] *concourantes*

[109] *tresse*, a notion defined here but not developed further in this work.

Commentary. Fabrics, mosaics and chessboards [a]

In her article, *The geometry of fabrics, mosaics, chessboards: Curious and useful mathematics* [238], Anne-Marie Décaillot wrote that in the second half of the nineteenth century, a group of mathematicians, driven by a common ambition of disseminating science to a wide audience, began to treat mathematical questions originating in concrete problems. To achieve their goal of making science popular, they favored simple visual representations. In the words of Décaillot:

> One of their favorite techniques was to employ the well-known common chessboard (*échiquier*). It suggested Édouard Lucas's geometry of fabrics (*géométrie des tissus*) with its connection to number-theoretic results. Then came Charles-Ange Laisant's construction of mosaics related to the representation of finite groups and crystallography. James Joseph Sylvester's analagmatic chessboards represented examples of recreational mathematics before their transformation into matrices attracted the attention of Jacques Hadamard.

In this same spirit, André Sainte-Laguë was influenced by the work of these nineteenth century mathematicians. We see this reflected in the chapter topics in the second half of this book. He continued carrying this message of making mathematics popular throughout his career by creating and employing visual representations for the public at large. We address this further in his biography later in the book.

---

[a] *Tissus, mosaïques, et échiquiers*

VIII   Chessboard problems

66. Chessboard. — The points of the plane with integer coordinates form a graph that we can consider as an infinite *chessboard*.[110] Many topological[111] questions can be reduced to the study of such graphs. Because these [questions] are usually unsolvable in general, we often consider bounded chessboards of n^2 cells[112], so the chessboard has a border.

Thus, Sylvester [197] has studied, under the name of *carrés* or *anallagmatic chessboards*[113] [185, 187, 195], square boards of black and white cells such that, for any two lines or any two columns, the number of *changes* of color is always equal to the number of *non-changes* of color. Their study is analogous to that of the formulas of decomposition of a product of sums of 4, 8, 16 , . . . squares into sums of 4, 8, 16 , . . . squares [178]. [114]

67. A large number of relations between the vertices or cells of a chessboard are represented by *chess pieces*.[115] This gives rise to the classical terminology.

A *king* can go from a cell (m, n) to one of the 8 cells $(m, n \pm 1)$, $(m \pm 1, n)$ or $(m \pm 1, n \pm 1)$. At each move, the king follows an edge of a regular graph[116] of degree 8.

A *rook* moves from (m, n) to (m, p) or (p, n). It follows the edges of a regular graph of degree 4.

A *bishop* moves from (m, n) to $(m \pm p, n \pm p)$. If we consider (m, n) as an *even (or white) cell*, or as an *odd (or black) cell*, depending on whether m and n have the

---

[110] S.-L. uses the term *échiquier indéfini* (indefinite chessboard). In this chapter, S.-L. relates much of what he has found in earlier literature on the number of various configurations of pieces on a chessboard, including formulas and tables that he quotes. We have translated these, and corrected some minor typos, but have not sought to verify any of the claims given in the text.

[111] *géométrie de situation*, in the terminology of S.-L.

[112] *cases*, in the terminology of S.-L.

[113] The term *anallagmatic*, from the Greek for "unchanging", refers to an object or structure that is not changed in form by inversion.

[114] See also [238, p. 191]

[115] Each according to the moves it may make.

[116] By claiming that the graph is regular, here and for all chess pieces below, S.-L. seems to be assuming that the chessboard is unbounded, or, more likely, assumes that the reader will understand that he ignores the boundary of a finite chessboard.

© Springer Nature Switzerland AG 2021
M. C. Golumbic, A. Sainte-Laguë, *The Zeroth Book of Graph Theory*,
Lecture Notes in Mathematics 2261, https://doi.org/10.1007/978-3-030-61420-1_9

same parity, we notice that the bishop moves along cells of the same parity. Turning the chessboard through 45°, we see that the movement of a bishop is identical to that of a rook. Sometimes one may consider a *half-bishop* [196] that moves from (m, n) to $(m \pm p, n \pm p)$, where the signs can be either the same or opposite.[117]

A *queen* is allowed to move both as a rook and as a bishop. She uses the edges of a graph of degree 8. A *half-queen* is allowed to move as a rook and as a half-bishop [196].

A *knight* moves from (m, n) to $(m \pm 1, n \pm 2)$ or $(m \pm 2, n \pm 1)$. In each such *jump*, it moves along the edges of a regular graph of degree 8. An *amazon* is allowed to move as a knight and as a queen [196].[118]

We may consider an infinite chessboard obtained by juxtaposing chessboards with n^2 cells. Each cell with coordinates $(\alpha n + p, \beta n + q)$ is then replaced by a *congruence cell* (p, q) (modulo n), in one of the chessboards of n^2 cells—generally the one in which p and q range from 0 to $n - 1$. Under this convention, a queen placed in the cell (p, q) can reach, or *commands*, any cell with abscissa p or ordinate q, and moreover the cells which, when shifted to the congruent cells of the primitive chessboard, are situated there on the primitive parallels to the diagonals, and also the segments of other parallels. A *great queen* is a queen that commands simultaneously all these cells of a chessboard with n^2 cells [196].

Commentary. An infinite chessboard from medieval Portugal.

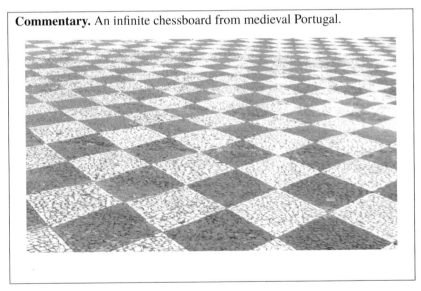

Image: common street tiles, 2020 – © Martin Charles Golumbic

---

[117] *Alain Hertz comments:* What Sainte-Laguë means is that there are two kinds of half-bishops. For the first type, the half-bishop can move from (m, n) to $(m + p, n + p)$ or to $(m - p, n - p)$. For the second type, it can move from (m, n) to $(m + p, n - p)$ or to $(m - p, n + p)$.

[118] The half-bishop, half-queen, and amazon are examples of unorthodox or fairy chess pieces which differ in the way they move. There are more than one thousand such unconventional pieces that have been incorporated into chess variations since the Middle Ages.

68. Following Lucas [186], the number of distinct moves that a knight can make on a rectangular chessboard with pq cells is $8pq - 12p - 12q + 16$. If it can move from a cell (m, n) to $(m \pm r, n \pm s)$ or to $(m \pm s, n \pm r)$, then this number becomes $8pq - 4(p + q)(r + s) + 8rs$, which must be divided by 2 if one of the numbers r, s, or $r - s$ is zero.

The moves of the other pieces can be expressed via the moves of a knight. This gives the following number of moves for a king:

$$8pq - 6p - 6q + 4.$$

On a chessboard of n^2 cells, this number becomes $4(2n - 1)(n - 1)$.

The number of moves for a rook is $2n^2(n-1)$; that for the two bishops, considered together, is $\frac{1}{3}2n(n - 1)(2n - 1)$; that for a queen is, $\frac{1}{3}2n(n - 1)(5n - 1)$.[119]

69. **Problems on rooks.** — The placement of n *pairwise non-attacking*[120] rooks is given by permutations, the question that we have already studied and to which we return in (§71). More generally, for a chessboard with pq cells, Lucas [186] established the following formula, in which E_r is the number of placements of r rooks:

$$rE_r = (p - r + 1)(q - r + 1)E_{r-1}.$$

from which we deduce that

$$E_r = r!C_p^r C_q^r.$$

Let us split the chessboard into two parts, and let T_r^s be the number of ways to place r pairwise non-attacking rooks so that s are in the first part, and $r - s$ are in the second. Then we have two formulas:

$$E_r = T_r^0 + T_r^1 + T_r^2 + \cdots + T_r^r$$

and

$$(p - r)(q - r)T_r^{s-1} = (r - s + 2)T_{r+1}^{s-1} + sT_{r+1}^s.$$

From this we deduce, for a chessboard of 64 cells, that the numbers f_r of different ways to place r bishops of the same color are given by the following table:

r	1	2	3	4	5	6	7
f_r	32	356	1704	3532	2816	632	16

The number F_n of different ways to place n pairwise non-attacking bishops, black or white, and on a chessboard of n^2 cells is given by

$$F_n = f_n + f_1 f_{n-1} + f_2 f_{n-2} + \cdots + f_{n-1} f_1 + f_n.$$

---

[119] *Myriam Preissmann comments:* The number of moves for a queen is equal to the sum of the number of moves for a rook plus the number of moves for the two kinds of bishops.

[120] *non en prise deux à deux*, in the terminology of S.-L.

This, according to Perrot [192], gives 22522960 solutions for $n = 8$.

70. Lucas [186] looked for the number Q_n of ways to place rooks [exactly n rooks on an $n \times n$ chessboard] so that no rook lies on a fixed diagonal. This gives the number of permutations where no element is in its natural position. We have the formula

$$Q_n = (n - 1)(Q_{n-1} + Q_{n-2}),$$

from which we deduce that

$$Q_n = nQ_{n-1} + (-1)^n,$$

and so

$$\frac{Q_n}{n!} = 1 - \frac{1}{1!} + \frac{1}{2!} - \frac{1}{3!} + \frac{1}{4!} - \frac{1}{5!} + \cdots + \frac{(-1)^n}{n!}.$$

This formula can be obtained from the number $P_n = n!$ of permutations of n elements, and can be written as

$$P_n = Q_n + C_n^1 Q_{n-1} + C_n^2 Q_{n-2} + \cdots + C_n^p Q_{n-p} + \cdots + Q_0,$$

where[121] $Q_{n-1}, Q_{n-2}, \ldots$ correspond to permutations with $1, 2, \ldots$ rooks on a chosen diagonal.[122] This can be expressed symbolically, after Neuberg [190], by[123]

$$P^{(n)} = (Q + 1)^{(n)}.$$

Because these rules of symbolic calculus can be applied to derivatives, Taylor's formula gives

$$(P + x)^{(n)} = (Q + 1 + x)^{(n)},$$

which is valid for all x. In particular, for $x = -1$, we get

$$Q^{(n)} = (P - 1)^{(n)}.$$

71. We find that the number D_n of ways to place n rooks symmetrically with respect to a diagonal, corresponds to *self-reciprocal* permutations (§60). It satisfies the identity,

$$D_n = D_{n-1} + (n - 1)D_{n-2},$$

and so

---

[121] *Corrected typo:* Q_{n-1}, Q_{n-2} instead of $Q_n - 1, Q_n - 2$.

[122] *Alain Hertz comments:* What Sainte-Laguë probably has in mind is that the number of permutations with x rooks on a chosen diagonal is equal to Q_{n-x} multiplied by the number C_n^x of ways to choose the x squares on the diagonal.

[123] The parenthesized superscript notation below is not defined by S.-L., but likely followed that of Joseph Jean Baptiste Neuberg [190]. For a biography of Neuberg, see http://mathshistory. st-andrews.ac.uk/Biographies/Neuberg.html

$$D_n = 1 + \frac{n(n-1)}{2} + \frac{n(n-1)(n-2)(n-3)}{2 \cdot 4} +$$
$$+ \frac{n(n-1)(n-2)(n-3)(n-4)(n-5)}{2 \cdot 4 \cdot 6} + \cdots .$$

The number B_n of ways to place n rooks symmetrically with respect to the two diagonals, satisfies the identity

$$B_{2n+1} = B_{2n} \qquad \text{and} \qquad B_{2n} = 2B_{2n-2} + (2n-2)B_{2n-4},$$

and this allows us to compute the terms B_n.

The number T_n of ways to place n rooks symmetrically with respect to a diagonal, and such that none of the rooks lies on that diagonal, is given by

$$T_{2n+1} = 0 \qquad \text{and} \qquad T_{2n} = (2n-1)T_{2n-2} = 1 \cdot 3 \cdot 5 \cdots (2n-1).$$

This leads to symbolic formulas

$$D^{(n)} = (T+1)^{(n)} \qquad \text{or} \qquad T^{(n)} = (D-1)^{(n)}.$$

If S_n is the number of the ways to place n rooks symmetrically with respect to the center of the chessboard, and such that none of the rooks lies on a diagonal, then

$$S_{2n+1} = 0 \quad \text{and} \quad S_{2n} = (2n-1)S_{2n-2} + (2n-2)S_{2n-4}.$$

Denoting by G_n the number of permutations symmetric with respect to the center of the chessboard, we then have

$$G_{2n+1} = G_{2n}, \qquad G^{(2n)} = (S_2+1)^{(n)}, \qquad G_{2n} = 2^n \cdot n! = 2^n P_n,$$

and also[124]

$$S^{(2n)} = (G_2-1)^{(n)} = (2P-1)^{(n)} = 2^n n! - \frac{n}{1!}2^{n-1}(n-1)! + \cdots + (-1)^n.$$

72. **Problems on queens.** — One of the most famous problems about queens is the "eight queens". This is a particular case of the following problem: *In how many ways can we put n queens on a chessboard of n^2 cells so that no two attack each other?* [156, 184, 194].[125]

Any solution to this problem can be represented by a permutation, and we see that *if a permutation occurs in this way, then so do its inverse permutation, its complement, and its reciprocal,* but in general this is not true of additive or circular

---

[124] *Corrected typo:* Above we changed $S^2 + 1$ to $S_2 + 1$, and below we changed $G^2 - 1$ to $G_2 - 1$.

[125] An extensive survey paper on the n-queens problem is [227]. In addition to discussing historical aspects of the original problem, and the mathematical literature, it includes extensions of the problem to other board topologies and dimensions.

permutations. By symmetry, each *solution-type* gives rise to 8, 4, or 2 distinct solutions, depending on the case.

Empirical studies have produced the number of solutions for small values of n. These are presented in the following table[126] [156]:

$n =$	1	2	3	4	5	6	7	8	9	10	11	12
non-symmetric solution-types				1	1							4
solution-types with 1 symmetry						1	2	1	4	3	12	18
solution-types with 2 symmetries					1		4	11	42	89	329	1744
total number of solution-types	1	0	0	1	2	1	6	12	46	92	341	1766
total number of solutions	1	0	0	2	10	4	40	92	352	724	2680	14032

It was established that, for $n \geq 3$, the problem of n queens always has at least one solution [156]. These can be written down as follows, when n is a multiple of 6 plus 0 or 4:

$$2, 4, 6, \ldots, n, 1, 3, 5, \ldots, n - 1,$$

and for n a multiple of 6 plus 2:

$$4, n - 2, n - 4, \ldots, 8, 6, n, 2, n - 1, 1, n - 5, n - 7, \ldots, 3, n - 3.$$

The case for n odd can be easily deduced from these.

---

[126] *Alain Hertz comments:* Note that for $n = 12$, the result is incorrect: there are 1787 solution-types and 14200 solutions (instead of 1766 and 14032).

73. The problem of placing n pairwise non-attacking *great queens* (§67) on a chess-board of n^2 cells has been considered in [193]. We obtain symmetric configurations which lead to *magic squares*.[127] Two of these configurations give rise to a third one via a *product* operation.

In this way, the following configurations on two chessboards, with 7^2 and 5^2 cells

$$1—5—2—6—3—7—4 \quad \text{and} \quad 3—1—4—2—5$$

give rise to the following configuration on a chessboard with 35^2 cells

$$3—1—4—2—5—23—21—24—22—25—8—6—9—7—10—$$
$$28—26—29—27—30—13—11—14—12—15—33—31—34—32—35—$$
$$18—16—19—17—20.$$

This example serves as a good illustration of the notion of the *product* operation.

The number of ways of placing p pairwise non-attacking queens on a chessboard of n^2 cells is for $p = 2$ [189]:

$$\tfrac{1}{6}n(n - 1)(n - 2)(3n - 1)$$

and for $p = 3$ [181, 191, 199]:

$$\tfrac{1}{12}(n - 1)(n - 3)(2n^4 - 12n^3 + 25n^2 - 14n + 1).$$

Is it possible to place 16 queens on a chessboard of 64 cells so that, on each row, column, or parallel to a diagonal, a queen attacks at most one other queen? This question assumes that there are two queens per row and per column, and leads to double permutations such as the following one, which gives a solution to the problem [156]:

$$5\ 4\ 2\ 1\ 2\ 1\ 3\ 4$$
$$8\ 6\ 6\ 7\ 3\ 7\ 5\ 8$$

---

[127] A *magic square* is an arrangement of the numbers from 1 to n^2 in an $n \times n$ matrix, with each number occurring exactly once, and such that the sum of the entries in each row, column, and diagonal is the same. Examples were known to the ancient civilizations of China, Persia, India, Arabia, and Greece. See, for example, the recent book by Jacques Sesiano [297].

Magic squares began to be studied mathematically in depth by Euler [135]; see, also De-launoy [172], Mansion [216], Fitting [241], John Conway, Simon Norton, and Alex Ryba [235] and Brian Hopkins and Robin Wilson [261]. A 4×4 magic square was featured in the 1514 engraving *Melencolia I* by Albrecht Dürer. They can be found in other recent works of art [274] and recreational mathematics [276].

Commentary. An illustration of Sainte-Laguë's example of 16 queens.

			Q		Q		
		Q		Q			
				Q		Q	
	Q						Q
Q						Q	
	Q	Q					
				Q		Q	
Q							Q

The following scheme gives the maximum number of queens, which is 11, such that every queen attacks two others and only them [155]; (the final column has no queens):

$$1\ 3\ 7\ 3\ 1\ 5\ 2$$
$$8\ 6\quad\ 7\quad\ 4$$

Finally, on a chessboard of 49 cells [156], we can place 49 queens of 7 different colors so that no two queens of the same color attack each other. If A, B, C, D, E, F, G is the order of colors of the queens in the first row, then the subsequent rows are deduced from this order by the circular permutations indicated by the colors of the first column: A, C, E, G, B, D, F.

Commentary. A board with 49 queens of 7 different colors with those of the same color being pairwise non-attacking.

$$A\ B\ C\ D\ E\ F\ G$$
$$C\ D\ E\ F\ G\ A\ B$$
$$E\ F\ G\ A\ B\ C\ D$$
$$G\ A\ B\ C\ D\ E\ F$$
$$B\ C\ D\ E\ F\ G\ A$$
$$D\ E\ F\ G\ A\ B\ C$$
$$F\ G\ A\ B\ C\ D\ E$$

74. *What is the minimum number p of queens that together attack all n^2 cells of a chessboard?* [154]. There are three different assumptions that one can adopt:

(a) the p queens, considered pairwise, can be [mutually] attacking or non-attacking;
(b) each of the p queens attacks at least one other queen;
(c) none of the p queens attacks any other.

Just as in the case of n queens (§72), we are interested in first finding *solution-types* that give rise, due to symmetries, to all other solutions. Following [154], we limit ourselves to the following table,[128] which lists numbers of solutions in the cases (a), (b), (c):

$n =$	1	2	3	4	5	6	7	8	9	10	11	12
p	1	1	1	2	3	3	4	5	5	5	5	6
(a) solution-types	1	1	1	3	37	1	13	638	?	?	1(?)	?
total number	1	4	1	12	186	4	86	4860	?	?	2(?)	?
p	0	2	2	2	3	4	4	5	5	?	6(?)	?
(b) solution-types	0	2	5	3	15	140(?)	5	56	?	?	?	?
total number	0	6	20	42	70	900(?)	22	352	?	?	?	?
p	1	1	1	3	3	4	4	5	5	5	5	?
(c) solution-types	1	1	1	2	2	17	1	91	?	?	1(?)	?
total number	1	4	1	16	16	120	8	728	?	?	2(?)	?

We also mention the following results:

7 queens, suitably placed in the central 5^2 cells, can attack the 13^2 cells of the chessboard;

9 queens, suitably placed among the central 7^2 cells, can attack the 17^2 cells of the chessboard, etc.

75. **Problems on knights.** — We can ask questions, similar to the preceding ones, about knights. In particular, we can ask how many knights do we need to place on a chessboard in order to attack all cells [155].

On another question, for a chessboard of width 4 and length n, Désiré André [158] looked for the numbers of ways, P_n, Q_n, R_n, S_n, in which a knight, which never goes back, can reach each of the 4 cells of the nth line starting from a given cell. Using the method of *complete arrangements*, he has shown that

$$P_n = Q_{n-2} + R_{n-1}, \qquad R_n = P_{n-1} + Q_{n-2} + S_{n-2}$$
$$Q_n = P_{n-2} + R_{n-2} + S_{n-1}, \; S_n = Q_{n-1} + R_{n-2}$$

From these, we obtain,

$$N_n = 2N_{n-2} + 2N_{n-4} + 4N_{n-5} + 2N_{n-6} - N_{n-8},$$

---

[128] The table entries marked "?" are as given in the original manuscript.

where N is any one of the letters P, Q, R, S. This is a recurrence relation arising from the generating equation

$$(x + 1)(x^3 - x^2 - x - 1)(x^4 + 2x - 1) = 0.$$

Commentary. We have translated the expression *non en prise* as *pairwise non-attacking*, even when Sainte-Lagüe omits adding *deux à deux*. This is the common term used today for the *n*-queens problem and similar combinatorial chessboard problems for rooks, bishops, knights, etc. Each of these classical combinatorial problems is limited to a single type of piece—for example, placing *n* rooks or *n* great queens on a chessboard. Thus, there is a certain symmetry between a pair of chess pieces of the same type: either they attack each other, and are therefore under attack by each other, or not.

En prise, in French, literally means *can be taken*, and in the chess community it describes any piece under attack, including one that can be captured by the opponent through an exchange of pieces in such a way that its loss would be disadvantageous.[a] Sir Harry Golombek, author, editor, and three-times British chess champion, wrote in the 1981 paperback revised edition of his *Encyclopedia of Chess*[b]: "When a player places a piece where it may be captured, then he is said to put the piece *en prise*."

Thus, in the chess world, *en prise* is a more general notion than in our classical combinatorial problems, where attacking (or being under threat of capture) is restricted to identical types of chess pieces acting symmetrically. But the expression *non en prise* has its source there.

---

[a] See http://www.chesshistory.com/winter/extra/enprise.html

[b] Original hardcover, London 1977.

IX Knight's tour

76. Euler's problem. — Like the problem of the 8 queens, another [chessboard inspired] problem of topology[129] that has been studied by many authors, is the *knight's tour problem of Euler*[130] [200, 208, 209, 215, 220]: *How can we find all the chains [sets of moves] of a knight across a chessboard, so that it visits each cell once and only once? How many knight's tours are there?* [131] This problem has also been generalized to three dimensions [201].

With respect to the permitted movements of the knight, the cells of a chessboard form a graph with 168 edges and 64 vertices, of which—16 cells, *the central cells*, are of degree 8; 16 are of degree 6; 20 are of degree 4; 8 are of degree 3; and 4 are of degree 2. We look for *complete chains* of a knight, or similarly *complete cycles* if the chain is to end where it started. Such chains are called *knight's tours*, and such cycles are called *closed knight's tours*.[132] Some enthusiasts can obtain from 10 to 12 different solutions in an hour, and one of them, in 1860 [215], obtained from 48 to 50 solutions in an hour (§81).

77. At each move, a knight jumps from an even (or white) cell to an odd (or black) cell or vice versa [67]. We deduce, as did Euler [208, 209], that *in a closed knight's tour, the sum of the numbers in each row or column is even.*[133] It follows that *there exists no knight's tour on a chessboard of any form such that the difference between the numbers of black and white cells is not 1 or 0* [215], and also *there are no closed knight's tours on a chessboard, square or not, whose number of cells is odd.*

78. To write down solutions, Vandermonde [221] proposed to denote each cell of the chessboard by a fraction $\frac{x}{y}$ where x and y are its coordinates. It is also possible

---

[129] *géométrie de situation*

[130] *Robin Wilson comments:* The problem is over 800 years old, and was solved pre-Euler. Euler was the first to study it mathematically.

[131] The knight's tour problem, and the closed knight's tour problem, are instances of the more general Hamiltonian path and Hamiltonian cycle problem, but unlike the general Hamiltonian problems which are NP-complete, the knight's tour problems can be solved in linear time. An extensive annotated chronology of the knight's tour problem compiled by George Jelliss can be found at https://www.mayhematics.com/t/1h.htm

[132] *marches rentrantes*

[133] An example is given below in (§78).

© Springer Nature Switzerland AG 2021
M. C. Golumbic, A. Sainte-Laguë, *The Zeroth Book of Graph Theory*,
Lecture Notes in Mathematics 2261, https://doi.org/10.1007/978-3-030-61420-1_10

to denote a cell as (x, y). Finally, we can label the cells of the chain according to the order in which the knight passes through them. With the latter notation, the classical solution of Euler is

$$
\begin{array}{cccccccc}
58 & 23 & 62 & 15 & 64 & 21 & 54 & 13 \\
61 & 16 & 59 & 22 & 55 & 14 & 51 & 20 \\
24 & 57 & 10 & 63 & 18 & 49 & 12 & 53 \\
9 & 60 & 17 & 56 & 11 & 52 & 19 & 50 \\
34 & 25 & 36 & 7 & 40 & 27 & 48 & 5 \\
37 & 8 & 33 & 26 & 45 & 6 & 41 & 28 \\
32 & 35 & 2 & 39 & 30 & 43 & 4 & 47 \\
1 & 38 & 31 & 44 & 3 & 46 & 29 & 42
\end{array}
$$

79. Various methods. — De Moivre [206] emphasized the role of the *central block* of 16 cells, and tried to make the knight pass through the other 48 cells as much as possible, making it pass through the central block of 16 cells only in the end.

Euler [208, 209] first traced a random chain, as long as possible, and then tried to make it into a closed tour. Afterwards, he tried to adjoin the unused cells.

Bertrand [220] completed this method by showing that it is possible to obtain new solutions given a solution that has already been obtained.

80. Laquiere [214] studied the cycles that one can form with the edges of the graph, or *partial closed knight's tours*. He then tried to connect these pairwise. He also sought to obtain the symmetric partial tours.

In this regard, Lucas [215] remarked that a closed knight's tour cannot admit a vertical, horizontal, or diagonal axis of symmetry, but it may have a symmetry center. One concludes that, in general, a closed knight's tour gives rise to 16 other tours. If it is closed, then since one can choose its origin in 64 different ways, one derives from it 1023 other closed knight's tours.[134]

81. These symmetries motivated Euler [208, 209] and other authors, such as Roget [218, 219], to look for a solution of the problem on a *half-chessboard*, bounded by the central horizontal line. By symmetry, we duplicate[135] [the half-chessboard solution], and Euler has shown how we can glue together the symmetric tours. Flye Sainte-Marie [210] divided each half-chessboard into cells that consist of 16 cells of the chessboard, and deduced 31054144 solutions.

This is the largest lower bound on the total number of solutions that we know. The obvious upper bound, but most likely very far from being tight, is C_{168}^{63}, a number that has more than a hundred digits.

82. Modern methods. — An ingenious solution to the *problem of Euler* was given by Warnsdorff [222]. At each move, he put the knight on the cell from which it

---

[134] *Alain Hertz comments:* In total, there are 1024 (16×64) different but equivalent tours, and given one of them, the 1023 others can be derived by symmetry.

[135] *S.-L.: on double par symétrie*

commands the least possible number of unused cells. This rule, which gives non-symmetric and non-closed solutions, could not be justified [mathematically], but has always worked well in practice, and seems correct for any rectangular chessboard. Warnsdorff thought that if the knight has a choice between two or more cells, it can take either of them, but one or two cases where this choice matters have been noticed.

83. Another interesting method is one of Roget [218, 219]. Having divided the chessboard into 4 blocks of 16 cells each, just like Laquière [214] and Flye Sainte-Marie [210], he labeled the 4 by 4 cells in each block, creating 4 labeled groups of cells with the same letter, a vowel A, E or a consonant B, C, as in the scheme shown in Fig. 14.

B	E	A	C
A	C	B	E
E	B	C	A
C	A	E	B

Fig. 14.

Having labeled the other *blocks* of the chessboard in a similar fashion, we connect the cells that are labeled by the same letter. If, for example, we denote the four blocks by subscripts $11, 12, 21, 22$ and the rows occupied by A in one of the blocks, by superscripts $1, 2, 3, 4$, we will have the following closed tour:[136]

$$A_{11}^3 A_{11}^4 A_{11}^2 A_{11}^1 A_{12}^2 A_{12}^1 A_{12}^3 A_{12}^4 A_{22}^2 A_{22}^1 A_{22}^3 A_{22}^4 A_{21}^3 A_{21}^4 A_{21}^2 A_{21}^1.$$

---

[136] *Corrected typo:* The incorrect version in the original manuscript was:
$$A_{11}^3 A_{11}^4 A_{11}^2 A_{11}^1 A_{12}^1 A_{12}^2 A_{12}^3 A_{12}^4 A_{22}^1 A_{22}^2 A_{22}^3 A_{22}^4 A_{21}^3 A_{21}^4 A_{21}^2 A_{21}^1.$$

Alain Hertz comments: The (corrected version of the) knight's tour corresponds to the chessboard illustrated as follows:

B	E	A	C	B	E	A	C
A	C	B	E	A	C	B	E
E	B	C	A	E	B	C	A
C	A	E	B	C	A	E	B
B	E	A	C	B	E	A	C
A	C	B	E	A	C	B	E
E	B	C	A	E	B	C	A
C	A	E	B	C	A	E	B

Using these cycles, the author showed that it is possible to compose a knight's tour that starts at a given cell and ends at a different given cell of different parity. Jaenisch [213] has given a method that is essentially similar to that of Roget.

Commentary. Peter Mark Roget

Peter Mark Roget, an English physician, is remembered mostly for his *Thesaurus of English Words and Phrases* (1852), but he also invented a "log-log" slide rule for calculating the roots and powers of numbers. He was an avid chess player and solved the general open knight's tour problem [218]. See the article, *Peter Mark Roget and Chess*, by Edward Winter at https://www.chesshistory.com/winter/extra/roget.html.

84. Following Moon [217], let us now start just as de Moivre did, with a central block of 16 cells labeled by A, B, C, D and the cells around it by a, b, c, d, as indicated in Fig. 15.

a	b	c	d	a	b	c	d
c	d	a	b	c	d	a	b
b	a	A	B	C	D	d	c
d	c	C	D	A	B	b	a
a	b	B	A	D	C	c	d
c	d	D	C	B	A	a	b
b	a	d	c	b	a	d	c
d	c	b	a	d	c	b	a

Fig. 15.

Starting from a cell [labeled] a, we can follow one of the two cycles $aDbCdAcB$ or $aDcBdAbC$. We thus obtain various chains that can be composed into a unique chain, which in general is not a closed tour. Moon showed that we can, using this method, obtain a chain that uses given initial and final cells.

Other authors, such as Collinis [205], also used the central block of 16 cells and the border of 48 surrounding cells, but in a different way. Other decompositions, such as that of Frost [211], have been proposed.

85. **Knight's magic squares**. — We conclude this chapter on knight's tours by noting that various authors, and in particular Beverley [203] and Wenzelides [223], have looked for and obtained various solutions of the *problem of Euler* in which, by numbering the cells in the order they are passed through by the knight (§78), we obtain a magic square. The sum in each row or column, but not a diagonal, is constant and equal to 260.

Some of these solutions give closed tours, such as the following one, which we give as an example:

47 10 23 64 49 2 59 6
22 63 48 9 60 5 50 3
11 46 61 24 1 52 7 58
62 21 12 45 8 57 4 51
19 36 25 40 13 44 53 30
26 39 20 33 56 29 14 43
35 18 37 28 41 16 31 54
38 27 34 17 32 55 42 15

Commentary. Yehuda Miklaf, a renowned Jerusalem contemporary artist and bookbinder, created the gift box below—goat skin with an inlaid design based on the theme of a magic square of side eight (the knight's tour).

Reproduced here by permission of the artist.

X Conclusion

The study of graphs can be pursued in many different ways, and each of the notions defined may initiate new research. We have investigated, to the best of our ability, the complexity of issues that are raised, and the variety of methods that have been employed and are indispensable. The subject, as limited as it may appear at first, is in fact vast and seems quite difficult.[137]

The research of Lucas and Tarry relate immediately to the theory of networks and graphs, and other applications of more immediate utility could also be considered. There is further work already started on planar graphs, and practical applications of graphs, as well as many questions involving higher arithmetic[138], topology[139], and game theory.

---

[137] Conclusion adapted from [4].

[138] *arithmétique supérieure*

[139] *géométrie de situation*

© Springer Nature Switzerland AG 2021
M. C. Golumbic, A. Sainte-Laguë, *The Zeroth Book of Graph Theory*,
Lecture Notes in Mathematics 2261, https://doi.org/10.1007/978-3-030-61420-1_11

A short biography of André Sainte-Laguë

Martin Charles Golumbic

André Sainte-Laguë (1882–1950) is perhaps best known historically for his method of *parliamentary seat allocation*, which he introduced in a short note [280] published in 1910, when he was 28 years old.[140] Known today as the *Webster–Sainte-Laguë Method*, having been proposed in 1832 by the American statesman, Senator Daniel Webster, it was first adopted in 1842 for the proportional allocation of seats in the United States House of Representatives. The method is still used today in many countries around the world. See also Gallagher [244], Lijphart and Gibberd [269] and Renwick [277].

Within the mathematics community, Sainte-Laguë was a pioneer of new educational technologies for teaching, promoting areas of recreational mathematics, mathematical games, and making mathematics accessible and understandable to the general public. He was particularly interested in research into the applications of mathematics, and was a prolific writer. *Les Réseaux* (*ou Graphes*), published in 1926, was the first of the many books he published during the following 25 years.

Early years

Jean André Sainte-Laguë was born on April 20, 1882, in the village of Saint Martin Curton in southwest France, between Bordeaux and Toulouse. The son of teachers, he spent part of his youth in Haiti and returned to France to finish his studies at the Lycée de Bordeaux. After his military service, he entered the prestigious École Normale Supérieure in 1903, graduating in mathematics in 1906. From 1906 to 1914, he taught high-school mathematics—first in Evreux, then in Douai, and finally in Besançon. Taking on leadership roles throughout his career, Sainte-Laguë was secretary of the Association of [secondary school/lycée] Mathematics Professors in 1910, and vice-president of the Society of [secondary school/lycée] Professors Agrégés in 1914.

---

[140] An English translation, "Proportional representation and the method of least squares", can be found in [269, Appendix 2].

© Springer Nature Switzerland AG 2021
M. C. Golumbic, A. Sainte-Laguë, *The Zeroth Book of Graph Theory*,
Lecture Notes in Mathematics 2261, https://doi.org/10.1007/978-3-030-61420-1

André Sainte-Laguë in 1937

Image: courtesy of the Palais de la Découverte, Paris

World War I

According to his official biography [234], with the outbreak of World War I, Sainte-Laguë was mobilized on August 2, 1914, and joined the infantry at the front in January 1915. Promoted successively from sergeant, to second lieutenant, to lieutenant, he received the *Croix de Guerre avec Palmes*, and in 1916 the military honor *Chevalier de la Légion* (Knight of the Legion). He was wounded three times, the third time in Verdun on June 24, 1916, when his thigh bone was broken by shrapnel and operated on five times.

Declared 'unfit for all weapons', Sainte-Laguë took advantage of 'trench recreation time' and his stays in military hospitals, as he once described it, to pursue mathematical research on graphs and topology which he had begun while at the École Normale Supérieure, and would eventually lead to *Les Réseaux*, his doctoral thesis in mathematical sciences in 1924.

From 1917 to 1919, Sainte-Laguë worked on long-range shell studies in the Department of Inventions and at the laboratories at the École Normale Supérieure.

Resuming teaching and research

After World War I, Sainte-Laguë returned to teaching mathematics at the prestigious Lycée Pasteur de Neuilly-sur-Seine, while continuing his research and completing his doctorate in June 1924. Listed on his dissertation committee are the well-known French mathematicians Émile Picard (President), Émile Borel and Paul Montel (Examiners). In 1927, Sainte-Laguë was appointed as a Maître de Conférences at the Conservatoire National des Arts et Métiers (CNAM), the French National School of Arts and Engineering.

On February 23, 1938, André Sainte-Laguë was appointed to the CNAM chair *Mathématiques en vue des Applications* (Mathematics for Applications). Jérôme Chastenet de Géry [234] has written, (translated from the original French):

> His classes enjoyed considerable success, unequaled until then, and one of his lectures had up to 2500 listeners, forcing him to give it three times in the great 900-seat Paul Painlevé amphitheater at CNAM. His warm and loud voice filled the room, and his lectures were lively, fast, and clear.

> When teaching mathematics to practitioners, he insisted on the necessity of theoretical, numerical, and graphical exercises, but also on the need to distinguish between rigorous reasoning and that which is not, as well as the importance of correct language. "We must not respond to the students' desire to know only *recipes*." A forerunner in what may now called 'new educational technologies', from 1928 onwards, he also used films for his geometry lessons. Apart from the media used, his method was hardly outdated in 1994.[141]

Besides his research in pure and abstract mathematics, Sainte-Laguë was interested and active in the application of mathematics to many areas of science and engineering. He published several works with his colleague Antoine Magnan, professor at the College de France, on the aerodynamics and flights of birds, gliders, and planes, and an essay on the motion of fish. Sainte-Laguë also wrote about the symmetries of nature, the world of form, and authored the book, *From Man to Robot*.

His book, *Avec des Nombres et des Lignes: Récréations Mathématiques* [289], was written for the 1937 Paris Exposition Internationale de Arts et Techniques dans la Vie Moderne (World's Fair). Sainte-Laguë was entrusted by Émile Borel with the organization of the mathematics rooms in the *Palais de la Découverte* (Palace of Discovery), the science museum created for the 1937 Paris Exposition. You can visit it today at the Grand Palais in Paris, including the famous *Pi Room*.

In 2018, the museum opened a new permanent space "Informatique et sciences du numérique" (Computer science and digital science). According to the museum website, the exhibit "addresses the four fundamental elements of computing: data, algorithms, languages and machines," as well as "current research themes: big data, machine learning (related to artificial intelligence), networks (*les réseaux*), and robots."

In the 1937 brochure of the *Section des Mathématiques*, André Sainte-Laguë is listed as the secretary and organizer of the exhibition. The other committee members were Émile Borel (president), Paul Montel (vice-president), Raoul Bricard and

---

[141] and is even more so today.

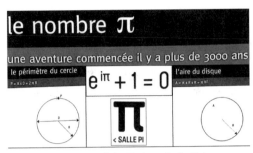

© Martin Charles Golumbic   Image: courtesy of the Palais de la Découverte

The Pi Room in 1937
Image: courtesy of the Palais de la
Découverte

The Pi Room in 2020
Image:
© Martin Charles Golumbic

Georges Darmois. In Salle 31 (Room 31), we have the following: "Three albums of curves by A. Sainte-Laguë, present 50 curves each to the visitors: algebraic, transcendental, and various others."

The official volume on the Palais de la Découverte from the 1937 Paris Exposition reports on p. 243 (translated from the original French):

> For those visitors who are fascinated by the research carried out by engineers, and who rave at the wonders of modern industry, the organizers presented some of the properties of the strength of materials in the simple and easily assimilated form of a film. The film, *De la similitude des longueurs et des vitesses* (Similarities between length and speed)[142] was written by André Sainte-Laguë.

---

[142] The film was directed by Jean Painlevé, a filmmaker well-known for his scientific documentaries, especially underwater fauna and sea animals. He directed more than two hundred science and nature films, including three additional films for the Palais de la Découverte: *La quatrième dimension* (The fourth dimension), *Images mathématiques de la lutte pour la vie* (The struggle for survival), and *Voyage dans le ciel* (Voyage to the sky). See https://jeanpainleve.org/ He was the only child of mathematician and twice prime-minister of France, Paul Painlevé, who appointed fellow mathematician and politician Émile Borel as Minister of Marine during his second (7 month) prime-ministry in 1925.

Image: courtesy of the Palais de la Découverte

World War II

In the spring of 1940, Nazi German forces occupied France. For André Sainte-Laguë the disaster of 1940 meant resistance against the occupation. He wrote: even if the chances of victory "would be only one in a thousand, it is necessary to hang on and help achieve the independence of France and that of intellect." Jérôme Chastenet de Géry [234] continues,

> He participated in the underground resistance in September 1940 and, known to be of a liberal mind, he was arrested by the Germans at his home in early October 1941 and imprisoned. But he knew how to remain discreet, and after many interrogations, the Germans, not finding evidence of his participation in clandestine activities, released him. When he resumed his class, he simply said to his pupils, "Gentlemen, I have been able to reflect for a long time, where I was, on the properties of the unicursal curves ...".

> It was not until the Liberation that they learned that he had been a leader of the Civil and Military Organization. He was decorated with the medal of the Resistance, and was appointed a member of the Provisional Consultative Assembly.

> One trait shows his solidarity and his altruism. During the Occupation, when his colleague and friend was deported, [Sainte-Laguë] replaced him to keep his office, and transmitted the friend's salary to his family.

The identity of this 'colleague and friend' was his co-author Guy Iliovici [265, 266]. The 1950 Fields Prize laureate and Bourbaki mathematician, Laurent Schwartz, wrote in his book "A Mathematician Grappling with His Century" [296]:

> A cousin of my mother's, Guy Iliovici, a professor of mathematics in the Lycée Saint-Louis, also remained in Paris and wore a yellow star. Because he was Jewish, he lost his job, but his colleague and close friend Sainte-Laguë gave him half of his salary. He was deported with his wife Suzanne and his daughter Janine; they never returned. [143]

Jérôme Chastenet de Géry continues,

---

[143] It was Jeannine's younger sister Nicole (1925-1942) who was deported with her parents and perished at Auschwitz. Jeannine was already married in 1937 and living elsewhere as were her older siblings Myriam and Michel.

A short biographical sketch of Guy Iliovici is included in the next section.

In addition to André Sainte-Laguë's professional, scientific, and patriotic activities, he invested a great deal in social and labor activism, particularly on behalf of the population whom he felt to be wedged between extremes. Among the founders of the Société des Agrégés in 1917, he was president and then honorary president. He was a leader of the *Confédération des Travailleurs Intellectuels* (Confederation of the Intellectual Workers) from the year of its foundation in 1921, and presided over it from 1929 to his death.[144] He was vice-president of the International Confederation of Intellectual Workers.

He was also deputy chairman of the *Comité National de Liaison et d'Action des Classes Moyennes* (National Liaison and Action Committee for the Middle Classes), vice-president of the National Economic Council, and in the latter title, he sat on the General Council of the Banque de France (but as a civil servant, he did not receive any director's fees).

André Sainte-Laguë died on January 18, 1950, after a short illness. Testimonials of sympathy flocked to the announcement of his death: *His life has been good, fiery, full of activity, while remaining simple and clear.*

The writings of Sainte-Laguë

The works and publications of André Sainte-Laguë are numerous, both in the scientific and technical literature and in popular and educational books. Among the additional references in the Bibliography, we include some of the most significant ones: [265]–[266], [270]–[272], [280]–[295]. Several comprehensive lists of his writings can be found at the websites:
https://www.idref.fr/032918658 and
http://data.bnf.fr/en/documents-by-rdt/12385472/70/page1

André Sainte-Laguë was a man of many dimensions—researcher, teacher, social activist, humanitarian. Most of all, he was an outstanding science communicator. His legacy is multifaceted, and thus will he be remembered.

---

[144] In her recent article [224], Anne-Sophie Aguilar writes, "In pre-War France, as central government cut back its commissions for artistic creation and the art market boomed, large numbers of artistic associations and societies joined the Confédération des Travailleurs Intellectuels (CTI), set up in March 1920 by a number of scholarly, literary and artistic associations to defend the interests of the as yet un-unionised intellectual professions."

MATHÉMATIQUES APPLIQUÉES

à l'usage des Ingénieurs, des Élèves-Ingénieurs et des Étudiants des Facultés des Sciences

PAR MM.

G. ILIOVICI

Professeur agrégé au Lycée Buffon,
Professeur à l'École spéciale
des Travaux Publics.

A. SAINTE-LAGUË

Professeur agrégé au Lycée Janson-de-Sailly.
Maître de conférences au Conservatoire
national des Arts et Métiers.

CALCUL INTÉGRAL : Aires planes et arcs. Volumes et aires de révolution. Masses et centres de gravité. Moments d'inertie.

CALCUL NUMÉRIQUE : Calcul numérique. Valeur numérique d'une fonction. Calcul numérique des séries. Racines d'une équation. Calcul intégral. Calcul graphique.

DEUX CENTS PROBLÈMES RÉSOLUS

PARIS

LIBRAIRIE DE L'ENSEIGNEMENT TECHNIQUE

LÉON EYROLLES, ÉDITEUR

3, Rue Thénard

1933

Tous droits réservés.

Title page of the book by Guy Iliovici and André Sainte-Laguë [265].

Image: courtesy of the Central Library of the Conservatoire National des Arts et Métiers (CNAM), Paris, [236]

Biography of Guy Ghidale Iliovici (1878–1942)

Martin Charles Golumbic

Guy Ghidale Iliovici was born on October 31, 1878 in Iasi (Jassy), Romania. He moved to France in 1902, becoming a naturalized citizen in January 1907. On 20 July 1910, he married Suzanne Gabrielle Trénel, daughter of Jacques Jacob Trénel and Rosa Lévy Trénel. They had four children, Myriam, Michel, Jeannine and Nicole. In late summer of 1942, Guy, Suzanne and Nicole (age 16) were deported by the Nazis and murdered at Auschwitz.

From 1907 until 1914, Iliovici taught mathematics at several schools: Lycée du Puy, Lycée de Vendôme, Lycée de Poitiers, Lycée de Besançon, and Lycée de Nancy. During World War I, he served three years at the front, first as a medic and then as a second lieutenant in artillery (SRS), until appointed Director of the SRS training center from August 1917 to February 1919, receiving the *Croix de Guerre*.

After World War I, Iliovici taught mathematics at the Lycée Michelet for one year before moving to the Lycée Carnot in 1920, succeeding Sainte-Laguë who had taught there part-time the previous year, and Professor Chollet. In 1926, Iliovici moved to the Lycée Buffon, where he remained until "retirement".[145] Guy Iliovici is mentioned in Lycée Buffon publications[146] as being among the *Professeurs du Lycée Buffon* honored as part of the Resistance and who died during World War II. On the contrary, his name is also listed the infamous book, *Ouvrages littéraires français non désirables* (1942), [*Unwanted literary works in France*] under the category of French-speaking Jewish authors.

Besides his books with André Sainte-Laguë [265, 266], Guy Iliovici was co-author of a book with Paul Robert [264], and a number of research papers in the literature, including [262, 263, 267].

Records indicate that Guy Iliovici had four siblings, Bernard, Claire, Rose (Rebecca) and Albert (Avram David). Albert Iliovici (1877-1965), trained as an electrical engineer, was a professor at the School of Public Works [l'École des Travaux

---

[145] Laurent Schwartz [296] wrote that Guy Iliovici (also) taught at the Lycée Saint-Louis, but we have not been able to confirm this.

[146] http://buffon.org/accueil.php?a=page201540 and http://buffon.org/accueil.php?a=page201520/Les-Morts-pour-la-France-1939-1945

© Springer Nature Switzerland AG 2021
M. C. Golumbic, A. Sainte-Laguë, *The Zeroth Book of Graph Theory*,
Lecture Notes in Mathematics 2261, https://doi.org/10.1007/978-3-030-61420-1

Publics]. He was awarded the Henri de Parville Prize on December 13, 1948 by the Academy of Sciences for his work on electrical measurements.

Suzanne Gabrielle Trénel Iliovici, wife of Guy Iliovici, came from a distinguished family. Her paternal grandfather, Rabbi Isaac Léon Trénel was the head of the *Séminaire Israélite de France*, now known as the *École Rabbinique de France*, during (1856-1890). Under his direction, the school was transferred from Metz to Paris by imperial decree signed by the Empress Eugénie on July 1, 1859.

Two of Suzanne's paternal uncles by marriage to daughters of Rabbi Isaac Trénel, Louise-Anna and Marianne, were mathematician Jacques Hadamard and Rabbi Simon Debré, spiritual leader of the Neuilly-sur-Seine synagogue from 1888 to 1939. Simon and Marianne's youngest son (Marc) Germain Debré was a well-known French architect; he enlarged and created a new facade for the synagogue in Neuilly-sur-Seine in 1937 with Julien Hirschal. Their middle son, Jacques Léon Debré, was an engineer and received the Chevalier de l'ordre du Mérite postal in 1954. Their oldest son, physician Robert Debré is considered one of the creators of modern pediatrics. The pediatric hospital, *l'Hôpital Robert-Debré* in Paris, has been named after him. His son, Michel Debré, was a former Prime Minister of France. Simon and Marianne's daughter Claire was the mother of mathematician Laurent Schwartz. These are some of Suzanne Iliovici's cousins.

Suzanne's older brother, Georges Henri Trénel (1886–1916), was killed in battle during World War I. [147] Her younger brother, Jean Trénel (1895–1943), was murdered by the Nazis at the Sobibor concentration camp.

---

[147] http://tadoukoz.net/picture.php?/29704/categories

Bibliography

The following bibliography[148] contains works that have historical interest. On the other hand, certain questions, such as the problem of queens, knight's tour, etc., require a more extensive bibliography than the one we give here. For these, one can refer to the classical treatises of Lucas [2], Rouse-Ball [3, 7], and especially Ahrens [1].

GENERAL WORKS

[1] AHRENS, WILHELM: *Mathematische Unterhaltungen und Spiele*, [Mathematical Recreations and Games], I and II. Leipzig, (1910–1918)

[2] LUCAS, ÉDOUARD: *Récréations Mathématiques*, I, II, III, IV. Paris (1882–1894)

[3] ROUSE-BALL, WALTER WILLIAM: *Récréations Mathématiques*, I, II, III. (Translation by Fitz-Patrick) [French] Paris (1907–1909)

[4] SAINTE-LAGUË, ANDRÉ: Les Réseaux. Paris or Toulouse (1924)

[5] SAINTE-LAGUË, ANDRÉ: Unpublished manuscripts (1922–1924)

[6] AHRENS, WILHELM: Mathematische Spiele. *Encyklopädie der mathematischen Wissenschaften* **I**, 2-ème Partie, Cahier **8**, 1080–1093 (1902)

[7] ROUSE-BALL, WALTER WILLIAM: *Mathematical Recreations and Essays*, 1st ed., London (1892); [French] translation see [3]. [10 editions by the author and 3 later editions revised by Harold Scott MacDonald Coxeter.]

[8] LUCAS, ÉDOUARD: *L'Arithmétique Amusante*. Paris (1895)

[9] LUCAS, ÉDOUARD: *Théorie des Nombres*. Paris (1891)

[10] SAINTE-LAGUË, ANDRÉ: *Annales Faculté des Sciences Toulouse* **XV**, (1923; Published in 1924)

[11] SAINTE-LAGUË, ANDRÉ: *Bulletin des Sciences Mathématiques* **48**, Comptes Rendus (1924)

[12] SAINTE-LAGUË, ANDRÉ: *Comptes Rendus Acad. Sci.* **176**, 1202–1205 (30 April 1923)

---

[148] References [1]–[223] are revised from the original 1926 version. Some of these are not actually cited in the manuscript, but were intentionally included by Sainte-Laguë. A small number still contain errors that could not be fully corrected. Additional references [224]–[304] are provided for this 2020 annotated translation.

© Springer Nature Switzerland AG 2021
M. C. Golumbic, A. Sainte-Laguë, *The Zeroth Book of Graph Theory*,
Lecture Notes in Mathematics 2261, https://doi.org/10.1007/978-3-030-61420-1

I. Introduction and definitions

[13] Antoine, Louis: Sur l'homéomorphie de deux figures et de leurs voisinages. Thèse, Paris (1921)
http://www.numdam.org/issue/THESE_1921__28__1_0.pdf

[14] Arnoux, Gabriel: Arithmétique graphique. Paris (1906)

[15] Arnoux, Gabriel and Charles-Ange Laisant: Arithmétique graphique. *Association Française pour l'Avancement des Sciences* (1900)

[16] Cayley, Arthur: The theory of groups: graphical representation. *American J. Math.* **1**, 174–176 (1878)

[17] Cayley, Arthur: On the analytical forms called trees with applications to the theory of chemical combinations. *Reports British Association* **45**, 257–305 (1875)

[18] Clifford, William K.: Cited by Sylvester [38] and Gordan [27]

[19] Cremona, Luigi: Les figures réciproques en statique graphique. French translation, Paris (1885)

[20] Dehn, Max and Poul Heegaard: *Analysis situs. Encyklopädie der mathematischen Wissenschaften* **III**, 1-ère Partie, Cahier **1**, 153–220 (1907)

[21] Delannoy, Henri A.: Les arbres. *Bulletin de la Société Chimique* **XI**, 3-ème série, 239–248 (1894)

[22] de Polignac, C.: Remarques sur les notations d'éléments … *A. E.*, 37–42 (1884)

[23] de Polignac, C.: Formules et considérations diverses se rapportant à la théorie des ramifications. *Bulletin Société Mathématique de France* **VIII**, 120–124 (1880); **IX**, 30–42 (1881)

[24] Errera, Alfred: Du coloriage des cartes et de quelques questions d'analysis situs. Thèse, Bruxelles (1921) and Gauthier-Villars, Paris (1921)

[25] Euler, Leonhard: Solutio Problematis ad Geometriam situs pertinentes. *Mémoires Académie de Berlin* (1759)

[26] Faà di Bruno, Francesco: Cited by Sylvester [38]

[27] Gordan, Paul and W. Alexejeff: Ubereinstimmung der Formeln der Chemie und der Invariantentheorie. *Sitzungsberichte der Societät zu Erlangen* (1900) 1–38 and *Zeitschrift für Physikalische Chemie* **35**, 610–633 (1900)

[28] Hadamard, Jacques: Représentation symbolique du résultant de deux équations. *Procès-verbaux Société des Sciences Bordeaux* (1894–1895), p. 24.

[29] Hermite, Charles: Cited by Sylvester [38]

[30] Listing, Johann Benedict: Census räuml. Kompl. Göttingen (1862)

[31] Lucas [2] **IV**,pp. 51–55; 240.

[32] Lucas [9] pp. 102–120.

[33] Mauguin, Charles: La structure des cristaux. Paris (1924)

[34] Petersen, Julius: Die Theorie der Regulären Graphs. *Acta Mathematica* **XV**, 193–220 (1891)

[35] Pólya, George: *Archiv der Mathematik und Physik Leipzig*, **III**, Reihe XXIV, 4. Heft, 369–375 (1916)

[36] Sainte-Laguë [4] pp. 3–6 (1924)

[37] SAINTE-LAGUË [5]: Les permutations.

[38] SYLVESTER, JAMES JOSEPH: On an application of the new atomic theory to the graphical representation of the invariants and covariants of binary quantics, with three appendices. *Amer. J. Math.* **1**, 64–128; 238–240 (1878)

II. TREES
[17, 20, 21, 23, 24, 34] and:

[39] CAYLEY, ARTHUR: *Berichte der Deutschen Chem. Gesellschaft* **VIII**, 1056–1059 (1875)

[40] CAYLEY, ARTHUR: Cited by DEHN AND HEEGAARD [20] 174–175 (1907)

[41] CHUARD, JULES: Questions d'*analysis situs. Thèse, Rendiconti Circolo Matematico Palermo*, 185–224 (1922)

[42] DELANNOY, HENRI A.: Les arbres. *L'Intermédiaire des Mathématiciens* **1**, 72–74 (1894)

[43] FRIEDEL, CHARLES: *L'Intermédiaire des Mathématiciens* **1**, 6 (1894)

[44] JORDAN, CAMILLE: Sur les assemblages de lignes. *Journal für die reine und angewandte Mathematik* **LXX**, 185–190 (1869)

[45] MACMAHON, PERCY ALEXANDER: *Philosophical Magazine* **XL**, 153 (1895)

[46] SYLVESTER, JAMES JOSEPH: Cited by DE POLIGNAC [23]

[47] TÉBAY, SEPTIMUS: *Educational Times reprints* **XXX**, 81 (1878)

[48] TÉBAY, SEPTIMUS: Cited by LUCAS [2] **I**, p. 51 (1882)

III. CHAINS AND CYCLES
[20, 25] and:

[49] AUBRY, AUGUSTE: Note sur les permutations. *Enseignement Math.* **20**, 199–215 (1918)

[50] BOUTIN, A.: Disposition de dominos. *L'Intermédiaire des Mathématiciens* **9**, 291 (1902)

[51] BRUNEL, GEORGES: Recherches sur les réseaux. *Procès-verbaux Société des Sciences Bordeaux* **5**, 165–215 (1895)

[52] CLAUSEN, THOMAS: *Astronomische Nachrichten, n°494.*

[53] DE LA CAMPA, S.: Géométrie anamétrique. *L'Intermédiaire des Mathématiciens* **6**, 29 (1899)

[54] DELANNOY, HENRI A.: Cited by LUCAS [2] **IV**,p. 134 (1894)

[55] ERRERA, ALFRED: Un théorème sur les liaisons. *Comptes Rendus Acad. Sci.* **177**, 489–491 (1923)

[56] FITTING, FRIEDRICH: Communications joignant *n* points. *L'Intermédiaire des Mathématiciens* **5**, 243; 1908 (1898)

[57] FLEURY, M.: Cited by LUCAS [2] **IV**, p. 134 (1894)

[58] FLYE SAINTE-MARIE, CAMILLE: Points reliés. *L'Intermédiaire des Mathématiciens* **24**, 120 (1917)

[59] HATZIDAKIS, NIKOLAOS J.: Problème de communications. *L'Intermédiaire des Mathématiciens* **8**, 111–115 (1901)

[60] HIERHOLZER, CARL AND CHRISTIAN WIENER: Ueber die Möglichkeit, einen Linienzug ohne Wiederholung und ohne Unterbrechung zu umfahren. *Mathematische Annalen* **VI**, 30–32 (1873)

[61] JOLIVALD, ABBÉ PHILIPPE [alias 'Paul de Hijo']: Cited by LUCAS [9] p. 108.

[62] LAISANT, CHARLES-ANGE: Cited by Lucas [2] **IV**, p. 126.

[63] LEMOINE, ÉMILE MICHEL HYACINTHE: Figures tracées d'un seul trait. *A. E.* (1881), 175–180 (1881)

[64] LEMOINE, ÉMILE MICHEL HYACINTHE: Problème de communications. *L'Intermédiaire des Mathématiciens* **6**, 51 (1899)

[65] LEMOINE, ÉMILE MICHEL HYACINTHE: Routes ne se recoupant pas. *L'Intermédiaire des Mathématiciens* **8**, 6 (1901)

[66] LIMINON: Entrelacements. *L'Intermédiaire des Mathématiciens* **18**, 132 (1911)

[67] LISTING, JOHANN BENEDICT: Vorstudien zu Topologie. *Göttinger Studien* (1848)

[68] LUCAS [2]: **I**, p. 36, 37, 51, 96, 102; **IV**, p. 133–151.

[69] LUCAS [9], pp. 51, 107–109.

[70] MAURICE: Cited by LUCAS [9], p. 51.

[71] MÉTROD, G.: Réseaux à trois carrefours. *L'Intermédiaire des Mathématiciens* **24**, 104 (1917)

[72] NETTO, EUGEN: Cited by AUBRY [49], p. 207.

[73] REISS: Cited by LUCAS [2], **IV**, p. 108.

[74] SAINTE-LAGUË: [4], 17–37 (1924)

[75] SEBBAU [Sebban, H.]: Points reliés. *L'Intermédiaire des Mathématiciens* **23**, 220 (1916)

[76] STEINERT, O.: *Archiv der Mathematik und Physik Leipzig* **XIII**, 220 (1895)

[77] STEINITZ, ERNST: *Monatshefte Vienne* **VIII**, 293 (1897)

[78] TARRY, GASTON: Parcours d'un labyrinthe rentrant. *Association Française pour l'Avancement des Sciences*, 49–53 (1886)

[79] TARRY, GASTON: Le problème des labyrinthes. *Nouvelles Annales de Mathématiques* **XIV**, 187–190 (1895)

[80] TARRY, GASTON: Cited by LUCAS [2] **IV**, p. 241.

[81] THUE, AXEL: *Tidskr. fur Math.* **III**, 102 (1885)

[82] TRÉMAUX, CHARLES PIERRE: Cited by LUCAS [9], p. 103.

[83] WELSCH: Problème de dominos. *L'Intermédiaire des Mathématiciens* **17**, 273–277 (1910)

IV. REGULAR GRAPHS
[34] and:

[84] BRUNEL, GEORGES: Les réseaux. *Procès-verbaux Société des Sciences Bordeaux*, 1894–1895 p. 3 (1894–1895)

[85] BRUNEL, GEORGES: Configurations régulières. *Procès-verbaux Société des Sciences Bordeaux*, (Juin 1898)

[86] CUNNINGHAN, ALLAN: *Reports British Association* **83**, 398 (1913)

[87] FAUQUEMBERGUE, E.: **21**, 33 (1914)

[88] HILBERT, DAVID: Cited by PETERSEN [34]
[89] MEISNER, W.: Cited by CUNNINGHAN [86] and FAUQUEMBERGUE [87]
[90] SAINTE-LAGUË: [4] 7–37 (1924)
[91] WIERNSBERGER: Recherches sur les polygones réguliers. Thèse, Lyon (1904)

V. CUBIC GRAPHS
[34, 44] and:

[92] AHRENS, Wilhelm: *Mathematische Annalen* **XLI**, 315 (1897)
[93] BRAHANA, Henry Roy: A proof of Petersen's theorem. *Annals of Math.* (2) **19**, 59–63 (1917)
[94] CHUARD, Jean: Propriétés des réseaux cubiques tracés sur une sphère. *Comptes Rendus Acad. Sci.* **176**, 73–75 (8 janvier 1923)
[95] ERRERA, ALFRED: Une démonstration du théorème de Petersen. *Mathesis* **XXXVI**, 56–61 (1922) An erratum appeared in *Mathesis* **XXXVII**, 318 (1923)
[96] KIRCHHOFF, G.: *Poggendorf Annalen* **LXXII**, 498 (1847)
[97] PETERSEN, Julius: Note sur les graphes. *L'Intermédiaire des Mathématiciens* **5**, 225 (1898) The correct title of this paper is—Sur le théorème de Tait.
[98] SAINTE-LAGUË: [4] 43–49 (1924)
[99] SAINTE-LAGUË: Les réseaux unicursaux et bicursaux. *Comptes Rendus Acad. Sci.* **182**, 747–750 (1926)
[100] TAIT, PETER GUTHRIE: Note on a theorem in the geometry of position. *Trans. Royal Society Edinburgh* **XXIX**, 657–660 (1880)

VI. TABLEAUX
[22, 28, 41, 51] and:

[101] ALEXANDER, JAMES WADDELL: See VEBLEN [126]
[102] CHUARD, JEAN: See DUMAS [103]
[103] DUMAS, G. AND JEAN CHUARD: Sur les homologies de Poincaré. *Comptes Rendus Acad. Sci.* **171**, 1113–1116 (1920)
[104] GAND, ÉDOUARD: Archives industrielles. *Cours de tissage* **I**, Paris (1886)
[105] HALPHEN, G.: Intégrales définies et discontinuités. Cited by PERRIN [115]
[106] KŐNIG, DÉNES: Ueber Graphen und ihre Anwendung aus Determinantentheorie und Mengenlehre. *Mathematische Annalen* **LXXVII**, 453–465 (1916)
[107] LAISANT, CHARLES-ANGE: Discours d'ouverture. *Association Française pour l'Avancement des Sciences*, (1879)
[108] LAISANT, CHARLES-ANGE: Régions du plan et de l'espace. *Association Française pour l'Avancement des Sciences*, (1881), 71–76 (1881)
[109] LAISANT, CHARLES-ANGE: Remarques sur la théorie des régions et des aspects. *Bulletin Société Mathématique de France* **X**, 52–55 (1882)
[110] LUCAS [9]: pp. 109–114.
[111] LUCAS, ÉDOUARD: Loi géométrique du tissage. *Association Française pour l'Avancement des Sciences*, Clermont Ferrand (1876) [and] Sur la géométrie du tissage. *Association Française pour l'Avancement des Sciences*, Paris (1878)

[112] Lucas, Édouard: Application de l'arithmétique à la construction de l'armure des satins réguliers. G. Retaux, Paris (1867)

[113] Lucas, Édouard: Principii fondamentali della geometria dei tessuti. *Ingegneria Civile e Arte Industriali* **VI**, 104–111, 113–115 (1880) Translation from Italian to French, 20 years after his death, by A. Aubry and A. Gérardin: Les principes fondamentaux de la géométrie des tissus. *Association Française pour l'Avancement des Sciences* **40**, 72–88 (1911)

[114] Merlin, E.: Configurations. *Encyclopédie des Sciences Mathématiques*, **III**, vol. **2**, fasc. 1, p. 148

[115] Perrin, R.: Sur le problème des aspects. *Bulletin Société Mathématique de France* **X**, 103–127 (1882)

[116] Perrin, R.: Aspects et configurations. *L'Intermédiaire des Mathématiciens* **1**, question 27 (1894)

[117] Poincaré, Henri: *Analysis situs J. de l'École Polytechnique*, 2ème série, Cahier **1**, 1–123 (1895)

[118] Poincaré, Henri: Complément à l'*Analysis situs Rendiconti Circolo Matematico Palermo* **XIII**, 285–342 (1899)

[119] Poincaré, Henri: 2ème Complément. *Proc. London Math. Society* **XXXII**, 277–308 (1900)

[120] Poincaré, Henri: 5ème Complément. *Rendiconti Circolo Matematico Palermo* **XVII**, 45–110 (1904)

[121] Rouse-Ball [3, 7]: Les satins, III, p. 315.

[122] Sainte-Laguë [4]: pp. 50–61.

[123] Sainte-Laguë [5]: Configurations et tissus.

[124] Steinitz, E.: Konfiguration der Projektiven Geometrie. *Encyklopädie der mathematischen Wissenschaften* **III**, A. B., 5 a, 481–516 (1910)

[125] Veblen, Oswald: An application of modular équations in *analysis situs*. *Annals of Math.* **XIV**, 86–94 (1912–1913)

[126] Veblen, Oswald and James Waddell Alexander: Manifolds of *n* dimensions. *Annals of Math.* **XIV**, 163–178 (1912–1913)

VII. Hamiltonian graphs
[37, 49, 53, 66, 72] and:

[127] Aubry, Auguste: Note sur les permutations. *Enseignement Math.* pp.281–294 (1917)

[128] Aubry, Auguste: *Sphinx-Œdipe*, (June 1913)

[129] Bioche, Ch.: Circuits avec les dominos. *L'Intermédiaire des Mathématiciens* **1**, question 68, (1894)

[130] Bioche, Ch.: Sur les permutations polyédriques. *Bulletin Société Mathématique de France* **XXXIII**, 88–89 (1905)

[131] Boris: Lecture de permutations avec grille. *L'Intermédiaire des Mathématiciens* **11**, 39 (1904)

[132] Bourguet, J.: Sur les permutations de *n* objets. *Nouvelles Annales de Mathématiques*, p. 147 (1883)

[133] BROCARD, HENRI: Courbe formée par un fil. *L'Intermédiaire des Mathématiciens* **7**, 280 (1900)

[134] CAUCHY: Cited by AUBRY [127], p. 289 and by LUCAS [9], p. 79.

[135] EULER, LEONHARD: Recherches sur une nouvelle espèce de quarrés magiques. *Verhandelingen Wetcnschappen te Vlissingen* **IX**, 85–239 (1872)

[136] FLYE SAINTE-MARIE, CAMILLE: Circuit avec des dominos. *L'Intermédiaire des Mathématiciens* **1**, 264 (1894)

[137] JACOBI: Cited by AUBRY [127], p. 285.

[138] JOLIVALD, ABBÉ PHILIPPE [alias 'Paul de Hijo']: *L'Intermédiaire des Mathématiciens* **11**, 159 (1904)

[139] KRONECKER: *Monatsberichte-Académie de Berlin*, (March and August 1869; February 1878)

[140] LAISANT, CHARLES-ANGE: Problème de permutations. *Bulletin Société Mathématique de France* **XX**, 105 (1890-1891)

[141] LAISANT, CHARLES-ANGE: *Comptes Rendus Acad. Sci.*, (11 mai 1891)

[142] LUCAS, ÉDOUARD: [9], pp. 65–69, 120.

[143] MÉTROD, G.: Nombres consécutifs dans permutations. *L'Intermédiaire des Mathématiciens* **24**, 27 (1917)

[144] MÉTROD, G.: Inversion de permutations. *L'Intermédiaire des Mathématiciens* **24**, 78 (1917)

[145] MÉTROD, G.: Permutations et suite de nombres. *L'Intermédiaire des Mathématiciens* **24**, 78 (1917)

[146] NETTO, EUGEN AND H. VOGT: Analyse combinatoire. *Encyclopédie des Sciences Mathématiques*, vol. 1, fasc. 1, 64–78.

[147] RODRIGUES, O.: Cited by AUBRY [127], p. 286.

[148] SAINTE-LAGUË, ANDRÉ [4]: pp. 8–9, 39–40, 47–49, 89.

[149] SAINTE-LAGUË, ANDRÉ [5]: Réseaux de degré 4.

[150] SAINTE-LAGUË, ANDRÉ [5]: Permutations.

[151] SAINTE-LAGUË, ANDRÉ [5]: Problème de timbres-poste.

[152] SAINTE-LAGUË, ANDRÉ [5]: Points doubles des courbes.

[153] WEBER, HEINRICH MARTIN: *Traité d'algèbre supérieure*, pp. 339–346, Paris (1898) French translation by J. Griess of *Lehrbuch der Algebra*.

VIII. CHESSBOARDS[149]
[107, 111, 112] and:

[154] AHRENS [1]: Les cinq reines, I, pp. 285–318.

[155] AHRENS [1]: Le cavalier, I, pp. 311–312; II, pp. 354–360.

[156] AHRENS [1]: I, pp. 210–284, 227; II, pp. 290, 344.

[157] AHRENS, WILHELM: Erreur dans la liste des huit reines. *L'Intermédiaire des Mathématiciens* **8**, 88 (1901)

---

[149] We have corrected a number of errors in the original bibliography, but do not claim to have found all of them. References [167, 168, 176, 181, 193] appear to coincide—the correct reference is Pitrat [193] and is the only one cited in the text.

[158] ANDRÉ, DÉSIRÉ: Application des arrangements complets au problème du cavalier. *Bulletin Société Mathématique de France*, (1876-1877)

[159] ANDRÉ, DÉSIRÉ: Arrangements complets. *Bulletin Société Mathématique de France*, (1878-1879)

[160] ARNOUS DE RIVIÈRE: Parcours en une seule fois. *L'Intermédiaire des Mathématiciens* **11**, 166 (1904)

[161] AUBRY, Auguste: Dé cubique qui roule. *L'Intermédiaire des Mathématiciens* **10**, (1903)

[162] BERDELLÉ, C.: Trajet de $(0,0,0)$ à (n,n,n). *L'Intermédiaire des Mathématiciens* **6**, 227 (1899)

[163] BOUTIN, A.: Pions non en prise. *L'Intermédiaire des Mathématiciens* **8**, 82 (1901)

[164] BOUTIN, A.: Marche de la tour. *L'Intermédiaire des Mathématiciens* **8**, 153 (1901)

[165] BOUTIN, A.: Dé cubique qui roule. *L'Intermédiaire des Mathématiciens* **9**, (1902)

[166] BOUTIN, A.: Routes par les points d'un quadrillage. *L'Intermédiaire des Mathématiciens* **10**, 181 (1903)

[167] BRAD: Les huit reines. *L'Intermédiaire des Mathématiciens* **7**, 330 (1900)

[168] BROCARD, HENRI: Les huit reines. *L'Intermédiaire des Mathématiciens* **7**, 330 (1900)

[169] BROCARD, H.: Dé cubique qui roule. *L'Intermédiaire des Mathématiciens* **9**, (1902)

[170] CATALAN: *Nouvelle Correspondance de Mathématiques* **VI**, 141

[171] DELANNOY, HENRI A.: Emploi de l'échiquier pour la solution des problèmes arithmétiques. *Association Française pour l'Avancement des Sciences*, (1886)

[172] DELANNOY, HENRI A.: Carrés magiques et cavalier. *L'Intermédiaire des Mathématiciens* **8**, 129 (1901)

[173] DE POLIGNAC, C.: Cited by LAISANT [107].

[174] DE ROCQUIGNY, G.: Trajet de $(0,0,0)$ à (n,n,n). *L'Intermédiaire des Mathématiciens* **6**, 227 (1899)

[175] FLYE SAINTE-MARIE, CAMILLE: Route passant par les points d'un quadrillage. *L'Intermédiaire des Mathématiciens* **11**, 86 (1904)

[176] GUNTHER: Les huit reines. *L'Intermédiaire des Mathématiciens* **7**, 330 (1900)

[177] KOPFERMAN: Problème des reines. *L'Intermédiaire des Mathématiciens* **11**, 162 (1904)

[178] LAISANT, CHARLES-ANGE: Développement de certains produits algébriques. *Association Française pour l'Avancement des Sciences*, (1881)

[179] LAISANT, CHARLES-ANGE: Géométrie des quinconces. *Association Française pour l'Avancement des Sciences*, 219–235 (1887)

[180] LANDAU, E.: Ueber das Achtdamen problem. *Natur Wochenschrift*, **XI**, (August 1896)

[181] LANDAU, E.: Les huit reines. *L'Intermédiaire des Mathématiciens* **7**, 330 (1900)

[182] LAGUIÈRE, EMMANUEL M.: Cited by LAISANT [107]

[183] LAGUIÈRE, EMMANUEL M.: Géométrie de l'échiquier. Gauthier-Villars, Paris (1880)

[184] LUCAS [2]: **I**, pp. 59–86.

[185] LUCAS [2]: Échiquier anallagmatique. **II**, pp.113–119; **IV**, pp.233–239.

[186] LUCAS [9]: pp. 96–102, 211–223.

[187] LUCAS, ÉDOUARD: Carrés anallagmatiques. *Association Française pour l'Avancement des Sciences*, (1877)

[188] LUCAS, ÉDOUARD: Reine non en prise. *L'Intermédiaire des Mathématiciens* **1**, 67 (1894)

[189] MANTEL, M.: Sur les combinaisons d'éléments dispersés dans un plan. *Association Française pour l'Avancement des Sciences*, 171–175 (1883)

[190] NEUBERG, JOSEPH: *Mathesis* **1**, 25.

[191] PAULS, E.: *Deutsche Schachzeitung* **XXIX**, 261–263 (1874)

[192] PEROTT, M.: Fous non en prise. *Bulletin Société Mathématique de France* **XI**, 173–186 (1882-1883)

[193] PITRAT, HENRI: Les huit reines. *L'Intermédiaire des Mathématiciens* **7**, 330 (1900)

[194] ROUSE-BALL [3, 7]: Les huit reines, II, pp. 116–124.

[195] ROUSE-BALL [3, 7]: Échiquier anallagmatique, II, pp. 39–42.

[196] SAINTE-LAGUË [5]: Amazones et grandes reines.

[197] SYLVESTER, JAMES JOSEPH: *Educational Times reprints* **X**, 74, 76, 112 (1868); **XLV**, 127; **LV** 97–99.

[198] TARRY, H.: *Association Française pour l'Avancement des Sciences*, (1890)

[199] TARRY, H.: *L'Intermédiaire des Mathématiciens* **2**, 205 (1895)

IX. KNIGHT'S TOUR
[8, 172] and:

[200] AHRENS [1]: **I**,pp. 319–398.

[201] AHRENS [1]: Le cavalier dans l'espace. **I**, pp. 384–386.

[202] BERTRAND: Cited by EULER [208]

[203] BEVERLEY, W.: On the magic square of the knights march. *Philosophical Magazine* **XXXIII**, 101–105 (1848)

[204] BOUTIN, A.: Chaîne du cavalier. *L'Intermédiaire des Mathématiciens* **10**, (1903)

[205] COLLINI, COSIMO ALESSANDRO: Solution du problème du cavalier au jeu des échecs. Mannheim, (1773)

[206] DE MOIVRE: Cited by ROUSE-BALL [3, 7] **II**, p. 220.

[207] DE POLIGNAC, C.: Note sur la marche du cavalier dans un échiquier. *Bulletin Société Mathématique de France* **IX**, 17–24 (1881)

[208] EULER, LEONHARD: *Mémoires Académie de Berlin*, 310–377 (1766)

[209] EULER, LEONHARD: Solution d'une question curieuse qui ne paraît soumise à aucune. *Memoires de Berlin*, p. 310 (1759); *Commentationes Arithmeticæ collectæ Saint-Pétersbourg* **I**, 337–355 (1849)

[210] FLYE SAINTE-MARIE, CAMILLE: Note sur un problème relatif à la marche du
      cavalier sur l'échiquier. *Bulletin Société Mathématique de France* **V**, 144–150
      (1876–1877)

[211] FROST, ANDREW HOLLINGWORTH: On the knight's path. *Quarterly J. of Math.*
      14, 123–125 (1877)

[212] GUNTHER, M. S.: Cited by MANSION [216]

[213] JAENISCH: Traité des applications de l'analyse mathématique au jeu des
      échecs, **II** (1862)

[214] LAQUIÈRE: *Bulletin Société Mathématique de France* **VIII**, 82–102, 132–158
      (1880)

[215] LUCAS [2]: **IV**, 205–223 (1894)

[216] MANSION, P.: Sur les carrés magiques. *Nouvelle Correspondance de Mathé-
      matiques* **2**, 161–164, 193–201 (1876)

[217] MOON, R.: On the knight's move of chess. *Cambridge Math. J.* **3**, 233–236
      (1843)

[218] ROGET, Peter Mark: *Philosophical Magazine*, series 3, **16**, 305–309 (April
      1840)

[219] ROGET, Peter Mark: *Quarterly Journal of Math.* **14**, 354–359 (1877)

[220] ROUSE-BALL [3, 7]: II, pp. 219–233.

[221] VANDERMONDE, ALEXANDRE-THÉOPHILE: Remarque sur les problèmes de
      situation. L'histoire de l'Académie des Sciences pour 1771, pp. 556–574, Paris
      (1774)

[222] VARNSDORFF [WARNSDORF, H. C. VON]: Des Rösselsprunges einfachste und
      allgemeinste Lösung. Schmalkalden, (1823)

[223] WENZELIDES, CARL: Bemerkungen über den Rösselsprunges nebst 72 Dia-
      grammen. *Schachzeitung* **IV**, 44 (1849)

Additional references

Martin Charles Golumbic

Readers interested in modern graph theory can find many recent and excellent textbooks: BOLLOBÁS [231], BONDY AND MURTY [232], DIESTEL [239], GOLUMBIC [245], GROSS AND YELLEN [256], WEST [301], and WILSON [303].

BIBLIOGRAPHY (CONTINUED)

[224] AGUILAR, ANNE-SOPHIE: Les artistes et le syndicalisme intellectuel dans la France de l'entre-deux-guerres: Le cas de la Confédération des travailleurs intellectuels (CTI In: Garric, Jean-Philippe (ed.), *Les dimensions relationnelles de l'art Processus créatifs, mise en valeur, action politique*, pp. 173–195. Éditions de la Sorbonne, Paris (2018)

[225] APPEL, KENNETH AND WOLFGANG HAKEN: Every planar map is four colorable. *Bull. Amer. Math. Soc.* **82**, 711–712 (1976)

[226] BEAUVILLE, ARNAUD: Géométrie des tissus [d'après S. S. Chern et P. A. Griffiths]. In: *Séminaire Bourbaki, 1978/79, Lecture Notes in Mathematics* **770**, 103–119, Springer, Berlin (1980)

[227] BELL, JORDAN AND BRETT STEVENS: A survey of known results and research areas for *n*-queens. *Discrete Mathematics* **309**, 1–31 (2009)

[228] BERGE, CLAUDE: *Théorie des Graphes et ses Applications*. Wiley (1958); English translation: *The Theory of Graphs and its Applications*. Wiley (1964)

[229] BERGE, CLAUDE: *Graphes et Hypergraphes*. Dunod, Paris (1970); English translation: *Graphs and Hypergraphs*. North-Holland (1973)

[230] BIGGS, NORMAN, E. KEITH LLOYD, AND ROBIN J. WILSON: *Graph Theory: 1736–1936*. Clarendon Press, Oxford (1998)

[231] BOLLOBÁS, BÉLA: *Modern Graph Theory*. Springer (1998)

[232] BONDY, J. ADRIAN AND U. S. R. MURTY: *Graph Theory*. Springer (2008)

[233] BRUALDI, RICHARD A.: Matrices of zeros and ones with fixed row and column sum vectors. *Linear Algebra and its Applications* **33**, 159–231 (1980)

[234] CHASTENET DE GÉRY, JÉRÔME: Sainte-Laguë, André (1882–1950): Professeur de Mathématiques générales en vue des applications (1938-1950) In: Fontanon, Claudine et André Grelon (eds.), *Les Professeurs du Conservatoire National des Arts et Métiers, Histoire Biographique de l'Enseignement*. INRP, Paris (1994)

[235] CONWAY, JOHN, SIMON NORTON, AND ALEX RYBA: Frenicle's 880 Magic Squares. In: Beineke, Jennifer and Jason Rosenhouse (eds.), *The Mathematics of Various Entertaining Subjects*, Vol. 2, pp. 71–84. Princeton University Press (2017)

[236] CNAM: Inventaire du fonds André Sainte-Laguë. Central Library of the CNAM, Paris, 2017. Archive of biographical documents and papers, donated by his granddaughter Dominique Sainte-Laguë.

[237] DE FRAYSSEIX, HUBERT AND PATRICE OSSONA DE MENDEZ: Trémaux trees and planarity. *European Journal of Combinatorics* **33**, 279–293 (2012)

[238] DÉCAILLOT, ANNE-MARIE: Géometrie des tissus, mosaïques, échiquiers: Mathématiques curieuses et utiles. [The geometry of fabrics, mosaics, chessboards: Curious and useful mathematics.] *Revue d'histoire des mathématiques* **8**, 145–206 (2002)
http://archive.numdam.org/article/RHM_2002__8_2_145_0.pdf

[239] DIESTEL, REINHARD: *Graph Theory*, 5th edition. Springer (2017)

[240] ELLINGHAM, MARK N. AND JOSEPH D. HORTON: Non-Hamiltonian 3-connected cubic bipartite graphs. *Journal of Combinatorial Theory* B **34**, 350–353 (1983)

[241] FITTING, FRIEDRICH: Rein mathematische Behandlung des Problems der magischen Quadrate von 16 und von 64 Feldern. [Pure mathematical treatment of the problem of magic squares of 16 and 64 squares.] *Jahresbericht der Deuschen Mathematiker-Vereinigung* **40**, 177–199 (1931)

[242] FLEISCHNER, HERBERT: *Eulerian Graphs and Related Topics.* Volume 1, *Annals of Discrete Math.* **45**, Elsevier (1990); Volume 2, *Annals of Discrete Math.* **50**, Elsevier (1991)

[243] FLEURY, M.: Deux problèmes de géométrie de situation, *Journal de Mathématiques Élémentaires*, 2nd ser., **2**, 257–261 (1883)

[244] GALLAGHER, MICHAEL: Proportionality, disproportionality and electoral systems. *Electoral Studies* **10**, 33–51 (1991)

[245] GOLUMBIC, MARTIN CHARLES: *Algorithmic Graph Theory and Perfect Graphs.* Academic Press, New York (1980); Second edition: *Annals of Discrete Math.* **57**, Elsevier (2004)

[246] GOLUMBIC, MARTIN CHARLES AND ANN N. TRENK: *Tolerance Graphs.* Cambridge University Press (2004)

[247] GROPP, HARALD: Configurations and graphs. *Discrete Math.* **111**, 269–276 (1993)

[248] GROPP, HARALD: Poincaré and graph theory. *Philosophia Scientiae* **1**, 85–95 (1996)

[249] GROPP, HARALD: On tactical configurations, regular bipartite graphs, and (v, k, even)-designs. *Discrete Math.* **155**, 81–98 (1996)

[250] GROPP, HARALD: Configurations and graphs II. *Discrete Math.* **164**, 155–163 (1997)

[251] GROPP, HARALD: Configurations and their realization. *Discrete Math.* **174**, 137–151 (1997)

[252] GROPP, HARALD: On configurations and the book of Sainte-Laguë. *Discrete Math.* **191**, 91–99 (1998)

[253] GROPP, HARALD: On combinatorial papers of König and Steinitz. *Acta Applicandae Mathematica* **52**, 271–276 (1998)

[254] GROPP, HARALD: The development of notation in graph theory in different languages. In: E. Fuchs (ed.), *Mathematics throughout the ages*, pp. 238–243, Prometheus, Praha (2001)

[255] GROPP, HARALD: "Réseaux réguliers" or regular graphs–Georges Brunel as a French pioneer in graph theory. *Discrete Math.* **276**, 219–227 (2004)

[256] GROSS, JONATHAN AND JAY YELLEN: *Graph Theory and Its Applications*, second edition. CRC Press (2006)

[257] GRÜNBAUM, BRANKO AND GEOFFREY C. SHEPHARD: Satins and twills: an introduction to the geometry of fabrics. *Mathematics Magazine* **53**, 139–161 (1980)

[258] GRÜNBAUM, BRANKO AND GEOFFREY C. SHEPHARD: Geometry of fabrics. In: F. Holroyd and R. J. Wilson (eds.), *Geometrical Combinatorics*, pp. 77–97. Pitman (1984)

[259] HOLTON, D. A. AND J. SHEEHAN: *The Petersen graph*. Australian Mathematical Society Lecture Notes **7**, Cambridge University Press (1993)

[260] HOPKINS, BRIAN AND ROBIN J. WILSON: The truth about Königsberg. *The College Mathematics Journal* **35**, 198–207 (2004); also in *Leonhard Euler: Life, Work and Legacy*, Robert E. Bradley and C. Edward Sandifer (eds.), *Leonhard Euler: Life, Work and Legacy*, pp. 409–420, Elsevier (2007)

[261] HOPKINS, BRIAN AND ROBIN J. WILSON: Euler's science of combinations. In: Robert E. Bradley and C. Edward Sandifer (eds.), *Leonhard Euler: Life, Work and Legacy*, pp. 395–408, Elsevier (2007)

[262] ILIOVICI, GUY: Sections sphériques d'un tore. *Nouvelles Annales de Mathématiques*, Series 6, Volume **2**, 1–2 (1927)
http://www.numdam.org/item/NAM_1927_6_2__1_0

[263] ILIOVICI, GUY: Quelques remarques sur les déterminants et les matrices. *Nouvelles Annales de Mathématiques*, Series 6, Volume **2**, 129–136 (1927)
http://www.numdam.org/item/NAM_1927_6_2__129_0

[264] ILIOVICI, GUY AND PAUL ROBERT: *Géométrie à l'Usage des Élèves de la Classe de Mathématiques, des Candidats aux Grandes Écoles (Saint-Cyr, Institut Agronomique, etc.) et des Élèves des Écoles Normales Supérieures de l'Enseignement Primaire*. Eyrolles, Paris (1937)

[265] ILIOVICI, GUY AND ANDRÉ SAINTE-LAGUË: *Mathématiques Appliquées à l'Usage des Ingénieurs, des Élèves-ingénieurs et des Étudiants des Facultés des Sciences*. Eyrolles, Paris (1933)

[266] ILIOVICI, GUY AND ANDRÉ SAINTE-LAGUË: *Cours d'Algèbre et d'Analyse à l'Usage des Élèves de Mathématiques Spéciales*, I et II. Eyrolles, Paris (1933)

[267] ILIOVICI, GUY AND G. WEILL: Quelques remarques de géométrie élémentaire sur les coniques considérées comme enveloppes de droites. *Nouvelles Annales de Mathématiques*, Series 6, Volume **1**, 65–67 (1925)
http://www.numdam.org/item/NAM_1925_6_1__65_0/

[268] KŐNIG, DÉNES: *Theorie der endlichen und unendlichen Graphen*. Akademische Verlagsgesellschaft, Leipzig (1936); Translated from German by Richard McCoart, *Theory of Finite and Infinite Graphs*. Birkhäuser (1990)

[269] LIJPHART, AREND AND ROBERT W. GIBBERD: Thresholds and payoffs in list systems of proportional representation. *European Journal of Political Research* **5**, 219–244 (1977)

[270] MAGNAN, ANTOINE AND ANDRÉ SAINTE-LAGUË: Étude des trajectoires et des qualités aérodynamiques d'un avion par l'emploi d 'un appareil cinématographique de bord. Gauthier-Villars, Paris (1932)

[271] MAGNAN, ANTOINE AND ANDRÉ SAINTE-LAGUË: *Le vol au point fixe*. *Actualités Scientifiques et Industrielles* **60**. Hermann, Paris (1933)

[272] MAGNAN, ANTOINE AND ANDRÉ SAINTE-LAGUË: *De Quelques Méthodes en Morphologie*. Masson, Paris (1933)

[273] MULDER, HENRY MARTYN: Julius Petersen's theory of regular graphs. *Discrete Math.* **100**, 157–175 (1992)

[274] NOWLAN, ROBERT A.: Magic squares and art. In: *Masters of Mathematics*, pp. 545–566. Sense Publishers, Springer (2017)

[275] PETERSEN, JULIUS: Sur le théorème de Tait. *L'Intermédiaire des Mathématiciens* **5**, 225–227 (1898)

[276] PHILIP, MORRIS: *Magic Squares: With Many New Additional Properties*. Philip Knitting Mills, New York (1986)

[277] RENWICK, ALAN: Electoral disproportionality: What is it and how should we measure it? (June 29, 2015)
Available at https://blogs.reading.ac.uk/readingpolitics/2015/06/29/electoral-disproportionality-what-is-it-and-how-should-we-measure-it/

[278] REYE, KARL THEODOR: *Geometrie der Lage*. [Geometry of Position.] (1876)

[279] REYE, KARL THEODOR: Das problem der configurationen. *Acta Math.* **1**, 93–96 (1882)

[280] SAINTE-LAGUË, ANDRÉ: La représentation proportionelle et la méthode des moindres carrés. *Académie des Sciences: Comptes Rendus Hebdomadaires* **151**, 377–378 (1910)

[281] SAINTE-LAGUË, ANDRÉ: *Notions de Mathématiques*. Hermann, Paris (1913)

[282] SAINTE-LAGUË, ANDRÉ: *Les Réseaux* (thèse Privat à Toulouse, and Hermann, Paris (1924)

[283] SAINTE-LAGUË, ANDRÉ: *Les Réseaux (ou Graphes)*. *Mémorial des Sciences Mathématiques*, No. 18. Gauthier-Villars, Paris (1926)

[284] SAINTE-LAGUË, ANDRÉ: *Notions de Géométrie Vectorielle*. Vuibert, Paris (1927)

[285] SAINTE-LAGUË, ANDRÉ: *Géométrie de Situation et Jeux*, *Mémorial des Sciences Mathématiques* No. 41. Gauthier-Villars, Paris (1929)
Available at http://eudml.org/doc/192569

[286] SAINTE-LAGUË, ANDRÉ: *Probabilités et Morphologie*. *Actualités Scientifiques et Industrielles* **36**. Hermann, Paris (1932)

[287] SAINTE-LAGUË, ANDRÉ: *Les Outils du Mathématicien*. Eyrolles, Paris (1933)

[288] SAINTE-LAGUË, ANDRÉ: *La Règle à Calcul*. Eyrolles, Paris (1934)

[289] SAINTE-LAGUË, ANDRÉ: *Avec des Nombres et des Lignes: Récréations Mathématiques*. Vuibert, Paris (1937)
Reprinted in 1994 and 2001 edited by André Deledicq and Claude Berge.

[290] SAINTE-LAGUË, ANDRÉ: *Du Connu à l'Inconnu*. Gallimard, Paris (1941)

[291] SAINTE-LAGUË, ANDRÉ: *Le Monde des Formes*. Fayard, Paris (1948)

[292] SAINTE-LAGUË, ANDRÉ: *Géométrie Descriptive et Géométrie Cotée*. Eyrolles, Paris (1948)

[293] SAINTE-LAGUË, ANDRÉ: *Algèbre, Analyse et Géométrie Analytique*. Eyrolles, Paris (1948)

[294] SAINTE-LAGUË, ANDRÉ: *Dessinons un Carré.* Vuibert, Paris (1950)

[295] SAINTE-LAGUË, ANDRÉ: *De l'Homme au Robot.* Hermann, Paris (1953)

[296] SCHWARTZ, LAURENT: *Un Mathématicien aux Prises avec le Siècle*, p. 215.
Odile Jacob, Paris (1997); English translation by Leila Schneps, *A Mathematician Grappling with His Century*, Birkhäuser, Boston (2001)

[297] SESIANO, JACQUES: *Magic Squares in the Tenth Century.* Springer (2017)

[298] TUTTE, WILLIAM T.: Les facteurs des graphes *Annals of Discrete Mathematics*
8, 1–5 (1980)

[299] WATE-MIZUNO, MITSUKO: Mathematical recreations of Dénes Kőnig and his
work on graph theory. *Historia Mathematica* **41**, 377–399 (2014)

[300] WIENER, CHRISTIAN: Ueber eine Aufgabe aus der Geometria situs. *Mathematische Annalen* **6**, 29–30 (1873)

[301] WEST, DOUGLAS B.: *Introduction to Graph Theory*, 2nd edition. Prentice Hall
(2001)

[302] WILSON, ROBIN J.: An Eulerian trail through Königsberg. *Journal of Graph
Theory* **10**, 265–275 (1986)

[303] WILSON, ROBIN J.: *Introduction to Graph Theory*, 5th edition. Prentice Hall
(2012)

[304] WILSON, ROBIN J.: *Four Colors Suffice: How the Map Problem Was Solved.*
Princeton University Press (2013)

Author index

© Springer Nature Switzerland AG 2021
M. C. Golumbic, A. Sainte-Laguë, *The Zeroth Book of Graph Theory*,
Lecture Notes in Mathematics 2261, https://doi.org/10.1007/978-3-030-61420-1

Geometric figures from SAINTE-LAGUË [285].

Image: courtesy of the Central Library of the Conservatoire National
des Arts et Métiers (CNAM), Paris, [236]

Index

© Springer Nature Switzerland AG 2021
M. C. Golumbic, A. Sainte-Laguë, *The Zeroth Book of Graph Theory*,
Lecture Notes in Mathematics 2261, https://doi.org/10.1007/978-3-030-61420-1

Glossary

French	English	Definition
aligné	traceable (has Hamiltonian chain)	(§5)
arbre	tree	(§6)
arête	edge	(§4)
articulation	cut vertex, articulation point	(§6)
associable, réseau	associable graph	(§8)
base (d'un arbre)	base of a tree (number of paths as above)	(§9)
bicubique	bipartite cubic	(§38)
bidiamétral	bidiametral	(§40)
cases	cells (of a chessboard)	(§66)
cerclés	circularly traceable (has Hamiltonian cycle)	(§5, §57)
chaîne	chain	(§5)
chaîne alternative	alternating chain	(§43)
chaîne orientée	oriented chain	(§5)
chaîne/circuit circulaire	simple chain/cycle (no repeating vertices)	(§5)
chaine/circuit complète	complete chain/cycle (hits all vertices)	(§5)
chemin, arête	edge	(§4)
chemin bicursal	bicursal edge, oriented edge with two traversal directions	(§5)
chemins jumelés	twin edges	(§36)
chemin pair/impair	even/odd edge	(§5)
chemin unicursal	unicursal edge, oriented edge with one traversal direction	(§5)
circuit	cycle	(§5)
circuit alternatif	alternating cycle	(§35)
classe d'un réseau	edge-chromatic number	(§7)
continent	interior faces, continent	(§7)
contour	contour	(§50)
côte	level	(§8)
croisement	crossing vertex	(§5)
cycle	oriented cycle	(§5)
degré d'un réseau	degree of a (regular) graph	(§7)
demi-échiquier	half-chessboard	(§81)
demi-chemin	half-edge	(§4)

M. C. Golumbic, A. Sainte-Laguë, *The Zeroth Book of Graph Theory*, Lecture Notes in Mathematics 2261, https://doi.org/10.1007/978-3-030-61420-1

French	English	Definitio
demi-circuits complets	complete half-cycles	(§25)
déterminant symmétrique gauche	left symmetric determinant	(§25)
dimensions	dimensions	(§8)
distance	distance	(§8)
élément d'un réseau polygonal	element of a polygonal graph	(§28)
entrelacement	interlacing (Eulerian chain or cycle)	(§5)
entrelacement fermé	Eulerian cycle	(§5)
entrelacement ouvert	Eulerian chain	(§5)
étoile	star	(§6)
extrémité	end, endpoint	(§4)
feuille	leaf component	(§6)
fini, réseau	finite graph	(§7)
genre	genus	(§7)
grande reine	great queen	(§67)
graphe	graph	(§4)
graphe aligné	traceable graph (admits complete simple chain)	(§5)
grosseur	maximal clique number	(§8)
hauteur	height	(§8)
impasse	pendant	(§4)
indice	index	(§7)
isthme	cut edge, isthmus, bridge	(§6)
jumelés	twins	(§35)
largeur	radius	(§8)
largeur d'un tissu	width of a semi-regular matrix	(§54)
longueur	diameter	(§8)
longueur d'un tissu	length of a semi-regular matrix	(§54)
magnitude (d'un arbre)	magnitude (twice the number of edges)	(§9)
marche du cavalier	knight's tour	(§76)
marche rentrante	closed knight's tour	(§76)
membre	leaf block, a subgraph without cut vertices	(§6)
mer	exterior faces, sea	(§7)
noeud (d'un arbre)	internal node	(§9)
non en prise	non-attacking	(§69)
noyau	maximal clique	(§8)
pas	move (in chess)	(§67)
permutation réciproque	reciprocal permutation	(§60)
polygone étoilé	star polygon	(§31)
polygones (T)	(T) polygons	(§31)
puissance	strength	(§8)
racine semi-primitive	semi-primitive root	(§32)
rameau (d'un arbre)	pendant edge (of a tree)	(§9)
rang d'un réseau	chromatic number	(§7)

French	English	Definition
réseau	graph	(§4)
réseau associable	associable graph	(§8)
réseau cerclé	Hamiltonian graph (has a complete simple cycle)	(§5)
réseau fini	finite graph	(§7)
réseau polygonal	polygonal graph	(§7)
réseau régulier	regular graph	(§7)
réseau rond	balanced graph	(§8)
réseau semi-carrelé	semi-tiled	(§48)
réseau semi-complet	self-complementary (a graph isomorphic to its complement)	(§8)
réseau simple/connexe	connected graph	(§5)
réseau sphérique	spherical graph	(§7)
réseau symétrique	symmetric graph	(§4)
réseaux jumelés	twin graphs	(§35)
rond, réseau	balanced graph	(§8)
semi-congruent	semi-congruent	(§32)
semi-exposant	semi-exponent	(§32)
semi-polygonal	semi-polygonal	(§7)
semi-premier	semi-prime	(§32)
singularité	singularity (multiple edge, pendant edge, loop, degree 2 vertex, cut vertex, cut edge)	(§7)
sommet	vertex	(§4)
sommet d'impasse	pendant vertex	(§4)
sommet de boucle	loop vertex	(§4)
sommet de croisement	crossing vertex	(§5)
sommet de passage	degree 2 vertex	(§5)
sommet isolé	isolated vertex	(§4)
sommet pair/impair	even/odd vertex	(§4)
sommets indiscernables	indiscernible vertices	(§4)
sommets symétriques	symmetric vertices	(§4)
tige	branch	(§12)
tissu	semi-regular matrix	(§52)
tissu mi-partie	mi-partite semi-regular matrix	(§54)
torique	toroidal	(§7)
trait	line	(§4)
trait (d'un arbre)	chain cover (of a tree)	(§9)
tresse	braid	(§65)

Acknowledgements

Throughout my professional career, I have been fortunate to work with hundreds of senior colleagues, postdocs, and research students on a variety of mathematical and computational topics. I have benefitted so much from our personal interaction over many years of academic collaboration that it would be impossible to fully express my appreciation to all of them. For this book on André Sainte-Laguë's 1926 original work, special thanks are due to Dmitry Sustretov who wrote an initial draft translation together with me during the academic year 2015–16 at the Hebrew University in Jerusalem. Michel Habib and Frederic Maffray encouraged me to pursue this project from the outset, and we had several lengthy talks about Sainte-Laguë's manuscript. Initial thoughts were also received from Bruno Courcelle.

Myriam Preissmann provided invaluable help on all aspects of the manuscript, reading the entire translation several times. We spent many hours at a time, deliberating the intentions of the author, understanding terminology which is often mathematically unclear, and the subtleties of presentation. Thanks to these detailed discussions, over a period of over two years, the process of altering, revising, and fine-tuning has greatly improved the translation. She also contributed significantly to this book with her extensive comments quoted throughout chapters IV, V and VI. Her dedicated scholarly assistance is greatly appreciated.

Alain Hertz provided many comments on Chapters VIII and IX that improved the translation further and clarified several aspects of the original manuscript. I thank him for all of this. Robin Wilson read the full manuscript, sharpening the exposition considerably with his unique ability as a prolific writer. He also contributed a number of historical notes.

I am pleased to acknowledge the support of the Caesarea Rothschild Institute of Computer Science at the University of Haifa, and its administrative coordinator Danielle Friedlander who assisted with the initial version of Sainte-Laguë's bibliography. I would also like to thank Abhiruk Lahiri for his help with the final formatting of the book, and for our many discussions on graph algorithms.

I benefitted greatly from having access to a large set of biographical documents recently donated to the central library of the Conservatoire National des Arts et Métiers (CNAM) by Dominique Sainte-Laguë, granddaughter of André Sainte-Laguë. I would like to thank Florence Desnoyers of the CNAM for her help in making this collection available to me in Paris. Finally, thanks are due to Angélique Durand from the Palais de la Découverte for making available to me several photographs and documents from the 1937 Paris exposition.

© Springer Nature Switzerland AG 2021
M. C. Golumbic, A. Sainte-Laguë, *The Zeroth Book of Graph Theory*,
Lecture Notes in Mathematics 2261, https://doi.org/10.1007/978-3-030-61420-1

I would like to thank Springer for 30 years of trust in me. This is my fourth book published with them, in addition to the journal *Annals of Mathematics and Artificial Intelligence* which I founded in 1990 and continue to run. I am happy to acknowledge my editors over the years, Lynn Montz, Welmoed Spahr, Ronan Nugent and Elena Griniari.

My deepest debt of gratitude, however, is to my wife, Lynn Pollak Golumbic, for her devotion and endless tolerance in understanding the quirks and craziness of this mathematician emersed in his research journeys. From the writing of my first book, *Algorithmic Graphs Theory and Perfect Graphs*, to the publication of this most recent one, she has given me the support and encouragement that allowed me to pursue my dreams.

Martin Charles Golumbic

Image: courtesy of the author

Martin Charles Golumbic is an algorithmic graph theorist, mathematician, and Emeritus Professor of Computer Science at the University of Haifa, Israel. He was a postdoctoral fellow in Paris with Claude Berge in 1976–77, and has maintained strong academic ties with colleagues from Paris, Montpellier, Clermont-Ferrand, Orsay, and Grenoble for over 40 years.

His translation and annotation of the original 1926 manuscript "Les Réseaux (ou Graphes)" by André Sainte-Laguë from French into English makes this valuable

© Springer Nature Switzerland AG 2021
M. C. Golumbic, A. Sainte-Laguë, *The Zeroth Book of Graph Theory*,
Lecture Notes in Mathematics 2261, https://doi.org/10.1007/978-3-030-61420-1

work more available as a historical reference for researchers in graph theory and combinatorial mathematics.

A native of the USA, Professor Golumbic received his Ph.D. degree in mathematics from Columbia University under the direction of Samuel Eilenberg (a non-French Bourbaki). He held positions at the Courant Institute of Mathematical Sciences of New York University, Bell Telephone Laboratories, the IBM Israel Scientific Center, and Bar-Ilan University, until moving to the University of Haifa as founding director of its Caesarea Rothschild Institute for Computer Science in 2000. He has also had visiting appointments at the Université de Paris, the Weizmann Institute of Science, École Polytechnique Fédérale de Lausanne, Universidade Federal do Rio de Janeiro, Rutgers University, University of Trento, Université de Montpellier, Columbia University, Hebrew University, IIT Kharagpur, Tsinghua University, and the University of New South Wales.

Martin Golumbic is the founding Editor-in-Chief of the journal *Annals of Mathematics and Artificial Intelligence*, and is a member of several journal editorial boards. He was elected as a Foundation Fellow of the Institute of Combinatorics and its Applications in 1995, has been a Fellow of the European Artificial Intelligence Society EurAI since 2005, and was elected to the Academia Europaea, honoris causa, in 2013. He is married to Lynn Pollak Golumbic, and they have four daughters, seven granddaughters and five grandsons to date.

Other books by Martin Charles Golumbic

Algorithmic Graph Theory and Perfect Graphs
      Academic Press (1980), Elsevier (2004)
Tolerance Graphs, (with Ann N. Trenk)
      Cambridge University Press (2004)
Fighting Terror Online: The Convergence of Security, Technology, and the Law
      Springer-Verlag (2008)

Edited books

Advances in Artificial Intelligence, Natural Language and Knowledge-based Systems, Springer-Verlag (1990)
Graph Theory, Combinatorics and Algorithms: Interdisciplinary Applications,
      (with Irith Ben-Arroyo Hartman), Springer-Verlag (2005)
Topics in Algorithmic Graph Theory,
      (with Lowell W. Beineke and Robin J. Wilson),
      Cambridge University Press (2021)

LECTURE NOTES IN MATHEMATICS  Springer

Editors in Chief: J.-M. Morel, B. Teissier;

Editorial Policy

1. Lecture Notes aim to report new developments in all areas of mathematics and their applications – quickly, informally and at a high level. Mathematical texts analysing new developments in modelling and numerical simulation are welcome.

   Manuscripts should be reasonably self-contained and rounded off. Thus they may, and often will, present not only results of the author but also related work by other people. They may be based on specialised lecture courses. Furthermore, the manuscripts should provide sufficient motivation, examples and applications. This clearly distinguishes Lecture Notes from journal articles or technical reports which normally are very concise. Articles intended for a journal but too long to be accepted by most journals, usually do not have this "lecture notes" character. For similar reasons it is unusual for doctoral theses to be accepted for the Lecture Notes series, though habilitation theses may be appropriate.

2. Besides monographs, multi-author manuscripts resulting from SUMMER SCHOOLS or similar INTENSIVE COURSES are welcome, provided their objective was held to present an active mathematical topic to an audience at the beginning or intermediate graduate level (a list of participants should be provided).

   The resulting manuscript should not be just a collection of course notes, but should require advance planning and coordination among the main lecturers. The subject matter should dictate the structure of the book. This structure should be motivated and explained in a scientific introduction, and the notation, references, index and formulation of results should be, if possible, unified by the editors. Each contribution should have an abstract and an introduction referring to the other contributions. In other words, more preparatory work must go into a multi-authored volume than simply assembling a disparate collection of papers, communicated at the event.

3. Manuscripts should be submitted either online at www.editorialmanager.com/lnm to Springer's mathematics editorial in Heidelberg, or electronically to one of the series editors. Authors should be aware that incomplete or insufficiently close-to-final manuscripts almost always result in longer refereeing times and nevertheless unclear referees' recommendations, making further refereeing of a final draft necessary. The strict minimum amount of material that will be considered should include a detailed outline describing the planned contents of each chapter, a bibliography and several sample chapters. Parallel submission of a manuscript to another publisher while under consideration for LNM is not acceptable and can lead to rejection.

4. In general, **monographs** will be sent out to at least 2 external referees for evaluation.

   A final decision to publish can be made only on the basis of the complete manuscript, however a refereeing process leading to a preliminary decision can be based on a pre-final or incomplete manuscript.

   Volume Editors of **multi-author works** are expected to arrange for the refereeing, to the usual scientific standards, of the individual contributions. If the resulting reports can be forwarded to the LNM Editorial Board, this is very helpful. If no reports are forwarded or if other questions remain unclear in respect of homogeneity etc, the series editors may wish to consult external referees for an overall evaluation of the volume.

5. Manuscripts should in general be submitted in English. Final manuscripts should contain at least 100 pages of mathematical text and should always include

   – a table of contents;
   – an informative introduction, with adequate motivation and perhaps some historical remarks: it should be accessible to a reader not intimately familiar with the topic treated;
   – a subject index: as a rule this is genuinely helpful for the reader.
   – For evaluation purposes, manuscripts should be submitted as pdf files.

6. Careful preparation of the manuscripts will help keep production time short besides ensuring satisfactory appearance of the finished book in print and online. After acceptance of the manuscript authors will be asked to prepare the final LaTeX source files (see LaTeX templates online: https://www.springer.com/gb/authors-editors/book-authors-editors/manuscriptpreparation/5636) plus the corresponding pdf- or zipped ps-file. The LaTeX source files are essential for producing the full-text online version of the book, see http://link.springer.com/bookseries/304 for the existing online volumes of LNM). The technical production of a Lecture Notes volume takes approximately 12 weeks. Additional instructions, if necessary, are available on request from lnm@springer.com.

7. Authors receive a total of 30 free copies of their volume and free access to their book on SpringerLink, but no royalties. They are entitled to a discount of 33.3 % on the price of Springer books purchased for their personal use, if ordering directly from Springer.

8. Commitment to publish is made by a *Publishing Agreement*; contributing authors of multiauthor books are requested to sign a *Consent to Publish form*. Springer-Verlag registers the copyright for each volume. Authors are free to reuse material contained in their LNM volumes in later publications: a brief written (or e-mail) request for formal permission is sufficient.

Addresses:
Professor Jean-Michel Morel, CMLA, École Normale Supérieure de Cachan, France
E-mail: moreljeanmichel@gmail.com

Professor Bernard Teissier, Equipe Géométrie et Dynamique,
Institut de Mathématiques de Jussieu – Paris Rive Gauche, Paris, France
E-mail: bernard.teissier@imj-prg.fr

Springer: Ute McCrory, Mathematics, Heidelberg, Germany,
E-mail: lnm@springer.com

Printed in the United States
By Bookmasters